1

WAMRA TECHNOPRISES, P.O. BOX 36665-00200, CITY SQUARE, NAIROBI, KENYA
wamra_technoprises_org@yahoo.com
2014

First published 2014
Set in Arial Bold 12
ISBN: 9966-7205-1-0

All Photographs used in this book have been produced and processed by the author except where acknowledged.

PUBLISHED BY
WAMRA TECHNOPRISES,
P.O. BOX 36665-00200, CITY SQUARE, NAIROBI
wamra_technoprises-_org@yahoo.com
TEL: +254722690956 / 738410345
2014

Dedication
This document is dedicated to my daughters
Michelle Achieng Amani,
Pendo Lorna Adhiambo and
Zawadi Faith Anyango

Table of Contents

SECTION 1: COMMUNITY MANAGEMENT OF WATER SERVICES

THE WATER AND SANITATION DECADE

The international water and sanitation decade (1980-1990) set out to focus international attention to the need for greater spending on drinking water and sanitation and to establish what contributes to the sustainability of services. Although considerable progress was made during this period some people are asking "why, after hundreds of millions of dollars of investment, do many of the world's poor still lack access to clean and basic sanitation?

Essay: Using evidence provided in your course material, describe why you think the above situation exists.

After the ten years of great dream about "water and sanitation for all", as a decade, there was an impressive yet insignificant progress towards achieving the objective. This could be attributed to many reasons and challenges as discussed below. These reasons include, but not necessarily limited to political, social, economic, financial, technical, and cultural.

First, the cultural element was never addressed to the level it deserved in water and sanitation (WATSAN) programs. The various religious societies and communities, such as the Muslims, have a strong very dimension to WATSAN. They view the use of water as a MUST in a mosque, and its use in a sanitation facility such as latrine as a second must. However, the design of these facilities in such set-ups is not just technical; there are many fine cultural elements which go with the water supply and sanitation in such environments. These were never taken into account when designing water schemes. Moreover, the water schemes were provided in some cases without due consultation with the primary beneficiaries, leading to operational and maintenance challenges.

Secondly, the financial requirements for WATSAN schemes were generally beyond reach of most institutions, compared to the demand. The capital costs were too high for any reasonable WATSAN schemes to be established. Even when donors made contributions, the money was never enough to ensure extension and rehabilitation of services. Thus immediately the projects attained their planned lifespan, they often collapsed. This left gaps on source of funds for operation and maintenance (O&M).

There were often no resources to rehabilitate or extend the service.

Thirdly, the technical services in relation to WATSAN only addressed hardware issues in the past. There was a lot of biasness, as the technical people were few, in great demand, and had to rush through paid projects to finish so as get other contracts. Here the private sector was used in a number of ways, and being profit-driven, they had no time for software aspects of the WATSAN projects. Moreover, the design of the projects often neglected the special needs of women and children, so that in a number of cases, they could not use these facilities and services even when they existed. For instance, the water pumps would be too heavy for women to handle, while the latrine holes would be too wide for young children to use. There was therefore a use-gap and questionable appropriateness of the facilities. As the facilities were less used, they were slowly vandalized so that people used the parts for other more immediate needs.

Fourth, the issue of spare parts for the water facilities led to even worse failure of water schemes. Lack of involvement of the primary stakeholders in the planning and design led to lack of clarity about where to get parts, how to fix them (if found), who to repair, who to operate and maintain etc. This led to ownership gap after project commissioning. Most O&M responsibilities (especially repairs) were left to a relevant government department after commissioning. The technical people were also unwilling to go to the field to do repairs of a project when they did not get inducements in form of allowances. This led to a technicality gap. Since the projects were largely technical, the failure was inevitable. In a number of cases, some foreign schemes were scaled up in developing countries prior to being locally tested to ascertain their suitability.

On the political front, there was poor distribution of the WATSAN facilities and services. The politically right areas were favored, especially in Africa where up to the end of that 1980s, there was single party era which greatly abused as a tool for settling political scores with personalities from communities sharing a geographical or racial background with the governments' perceived enemies. This was too clear in the case of Kenya, where the Broad Nyanza province was neglected by the government (pre and post- independence). The post-independence government gave Nyanza an economic embargo due to political differences between the heads of the government and an individual from Nyanza, whom the governments also viewed as a threat. This still continues. Due to lack of checks and balances, the governments often embezzled the project funds.

Further to political factors, willingness to charge issues has arisen in relation to user pay-approach in operation and maintenance. Politicians, being policy makers, and vote seekers at the same time, have been seriously compromised. They fear being voted out for bringing up unpopular policies of their constituents paying for services that the public think the government should provide free. In essence this is willingness to charge issue. This has been worsened by non-involvement of the voters / community members during project stages. Also lack of enough funds to expand the services, including upgrading and increasing levels of service, have been lacking.

From a broader political perspective, the international treaties governing the use of common boundary (international) waters such as Lake Victoria played a key role in compromising international community's ability to supply water to the wider Lake Victoria basin. In the Lake Victoria situation, a treaty was signed in 1929 between the Egyptian government and the colonial Kenya government baring the counties surrounding Lake Victoria from using the water of the lake on large scale. Due to this limitation in use of the lake waters, the governments of Uganda, Kenya and Tanzania have never taken any bold step towards protecting the lake from pollution. This is greatly compromising the ability of the lake communities to have water. In Kenya, the worst cases of water related diseases are recorded in Nyanza and western provinces, both of which inhabit Lake Victoria basin. This political sabotage of the East African communities has therefore compromised the ability of its citizens getting improved water supply.

In conclusion, various factors have contributed to low achievement in the water and sanitation growth despite heavy investment. Non-inclusive planning rates (hardware approach) first, flowed by Political factors, and lastly technological factors (inappropriateness).

REASONS FOR SANITATION LAG BEHIND WATER

Why is basic sanitation coverage falling behind progress made in water supply?

Water and sanitation (WATSAN) programs have always been integrated. The water supply part is largely communal / public and encompasses hardware and software aspects, both of which have been appropriately addresses to varying degrees, with the software aspect trailing. On the sanitation side, however, both the hard ware and soft ware issues have been given a raw deal. This is largely because whereas water supply is communal, and any one water facility may benefit many people in different ways, the sanitation facilities are largely individual. In essence, they affect largely the household where decisions on whether to have them or not, when, what type etc are purely household in nature. The sanitation issue is therefore likely to be relegated where choice between putting up a sanitation facility and a basic need (e.g. water etc) is to be made.

It has been easy to give water priority because with the climatic change, coupled with siltation of hitherto permanent local / community water resources (ponds), the latter have become increasingly insecure; some have dried up, leaving the communities with no reliable water source. It has therefore been easier for communities to pool resources together when need arises, to put up water facilities with or without external assistance. This has forced individual households to contribute (in cash or kind), as the facility is mandatory, as well as to fulfill their individual roles and responsibilities. However, with meager household income, the facilities which are purely home-based have been neglected. In cases where it has been made mandatory (with the sub chief inspecting, and punishing those without), there are cases where toilets / latrines are constructed, but never used. They are thus made for inspection purposes. This is attributed to the fact that views and approaches towards sanitation are highly diverse.

Secondly, sanitation is never a common priority, as it is influenced by individual cultural, economic, social and environmental setting. For instance, in places where people have been used to using the bushes, it cannot be easy to convince them to make sanitation facilities as long as the bushes still exist. Thirdly, there are many taboos associated with sanitation facility use. One example is whereby the in laws are not supposed to share a toilet (i.e., a son or daughter in law cannot share a toilet with a mother or father in law. This is applicable to the Luhya and Luo communities in Kenya, as well as the bulk of the Christian communities in Southern Sudan. This means that at any given time, one has to construct at least two latrines in a typical Luo / Luhya homestead in Kenya. Logistics therefore make toilet construction difficult.

Fourthly, sanitation requires a lot of software input to inform demand for it given peoples' diverse backgrounds and opinions about it. This could convince more to prioritize it, alongside hygiene practices. However, the twin combination of sanitation and hygiene are often neglected in the design of WATSAN projects. They demand too much from the technical people whose main aim, interest and strength lie in hardware. Thus the WATSAN projects often demand a social intervention, especially social marketing. Yet social scientists well groomed in the human science of extension never used to have a place in "technical programs" as it was imagined what was needed was the facility (hardware) with no need for follow-up, leave alone doing a baseline survey to determine the background demand for the facilities. In most cases, therefore, the facilities remained white elephant projects littering different sections of the country-side.

The order of planning and implementation of sanitation programs have always assumed this vital social need, which has literally made it supply driven. Most supply driven programs are never sustainable. The approach assumed there was simply a sanitation and hygiene or technological vacuum among the primary stakeholders. The latter were therefore assumed, and there was no participatory planning for the WATSAN services. Since the traditional networks, authorities and institutions felt belittled by being ignored, some even sabotaged the projects. In any case, the majority of communities feel better when an idea is introduced directly to them or through their leaders (credible institutions whose operations they know, and regard highly). This exclusive and lone-ranger approach was thus the major reason why sanitation programs tail behind water supply programs.

In essence, therefore, sanitation trails in WATSAN projects largely due to lack of social marketing and other aspects of software required to support it.

COMMUNITY CONCERNS, NEEDS, ASPIRATIONS AND PARTICIPATION

Essay: If we are to help people take charge of their lives, why is it so important to take into account the perspectives, needs and preferences of communities and households when designing water and sanitation interventions?

People perspectives give a strong indication of their interest, priorities and understanding and interpretation of the WATSAN interventions; Peoples needs need to be real needs, not perceived (imaginary needs); these guide us in identifying the priorities of the people, and what they would be most willing to get involved, if not fully participate in. Addressing peoples' needs impart demand drive into a project. This recognizes water and sanitation projects as economic goods to which the primary stakeholders would be most willing to pay for and maintain.

If a project cycle is systematically developed, participatory procedures and tools can be used to identify community needs by conducting baseline survey which sets the project foundation and provide a basis for project monitoring and evaluation. This forms the basis of the planning process, and provides the answers to the question: "Where are we now?" Once this question is honestly answered, then we can know what is left, once the community has stated where they want to be. Only thereafter is it possible to honestly decide HOW to reach the destination / intended result.

Through this participatory planning process, the community positively identifies its priorities, even in situations where cloudiness may have caused initial confusion on the best path to take. Thus addressing the real needs of the community renders them as owners of the project; project ownership by the community is the single most important tool in ensuring sustainability of the project. Sustainability means having the benefits associated with a project being felt and practiced long after the life of the project comes to an end. It is recorded that community-managed projects have the longest life. Thus identifying the community's real needs ensures project sustainability.

Therefore if we are to help people take charge of their lives, it so important to take into account community and household perspectives, needs and preferences when designing water and sanitation interventions, among others. Since a project is supposed to bring a positive change, the change should be one which the community identifies with; this can only happen if the project took in to account their real needs at the onset.

SUSTAINABILITY OF WATER PROJECTS

Essay: What has been learned about sustainability from the experience of agency managed and community managed projects?

Management means planning on the use of resources such as finances, personnel (human resource), facilities / hardware and time. It means making decisions on what to do, when, where, how, and why. In economic terms, a manager makes key decisions, and bears the consequences of such decisions. Thus it is possible for the community to pool together their resources. In Kenya, this is called Harambee. It discourages shyness even among those with no money, or generally deemed to be poor, as community accepts any kind and form of contribution- be it in form of skills (e.g. management, organizational and supervision), monetary, networking, labour, or in any form. Thus even those deemed to be poor normally feel at home, as they make small, useful contributions highly appreciated by all. Since the commitment involves sacrifice on the part of all, community members tend to protect the project and identify with it so intimately. It becomes a sense of pride, a living testimony, and a tangible result of what they can do. With the above in mind, a community managed project instills a sense of ownership from the very beginning. A community-owned project guarantees project sustainability.

A case in point is a Majango community/ TICH (Kisumu-based TRUST) partnership mabati/roof catchment tank project. After initial baseline survey prior to formalization of the partnership, participatory planning took place. The community identified water as their key problem. They made an action plan spanning 5 years, in which they designed a merry –go-round [project. In this, a member was to get members' collective contribution in form of Iron sheets for use in building semi-permanent houses. Once the first round was done, the second round focused on roof tank for each member, all of whom already had Iron roofed houses already. This cycle is ending in October, having left a permanent mark in the villagers' lives. TICH (initially a TRUST based in Kisumu city in Kenya, later gave rise to university,

12

GLUK), does not come in at all, except transporting material when requested, on its scheduled days of visiting the community on partnership days.

An agency managed project, on the other hand, does not have home-grown ideas, and planning is non-participatory. These are then most likely to be forced onto the community, who then can react by resenting and rejecting the project, however noble the idea might be. The foreign ideas may not be the community's priority, even if the issue addressed is a real problem there. A sense of ownership normally associated with inclusive planning with relevant stakeholders is lacking, as the primary stakeholders feel assumed and ignored. Such agency managed projects therefore often end as soon as funding stops, as there no momentum to keep them going. Examples are local catholic churches in Kenya. In Bondo district there is Nyangoma Catholic Church which though a big complex, is still known as Kodera, meaning the home of Odera. Odera was the local name given to the founding catholic Priest of Nyangoma church- as the closest to Auderaa which was the real name. The priest died more than 20 years ago. Similarly, Lwak Catholic Church is to date called Ka Ayot- to mean the home of Ayot. Ayot was the first and founding priest of Lwak mission.

ROLE OF GOVERNMENT IN WATER SUPPLY
Essay: What are the implications of the new role of government (from provider to facilitator of services)?
Traditionally, the government was supposed to be everything in a society, operating more as a jack of all trades. It was meant to play all the roles relating to development. This had a number of advantages, but many more disadvantages. Some of these included financial, technological and logistical limitations.

On the positive side, it was possible to point fingers at the government as a failure. However, it was not possible to point at any specific department or individual within the same government to take responsibility. This was attributed to lack of clear-cut roles and responsibilities among ministries and departments. Some roles overlapped, leading to no department doing them. The departments were also not accountable to anyone, as there was the bond of collective responsibility. Regular transfer of departments from ministry to ministry during cabinet reshuffles also caused more inefficiency. There was therefore complete lack of accountability and transparency.

13

The fact that the government was everything greatly limited operations of other alternative development partners, facilitators and providers who felt there was no role for them, yet no work was done successfully. Traditionally, international partners, who had some resources for development, passed the same to governments to use for implementing some programs. But the government, being a jack of all trades, was grossly overworked, became even more inefficient and ineffective. It was neither accountable to anybody, nor showed any sense of transparency. There was gross abuse of power. It was never able to offer the basic services to enable it account for the resources contributed by other development partners.

Regular evaluations of government programs after 5-year planning period usually showed very little on the ground. This trend made other development partners to think twice and rechannel their monies to institutions which could deliver. Thus in the 1980's, policy changes called special adjustment programs (SAPs) were proposed, and some literally forced on governments to take up. This is when the idea of development partnership was born, and it involved incorporating government, private sector, communities, intermediaries, parastatals, international institutions (UNDP, World Bank, IMF etc) alongside the government. The idea was to assign each group a role it was best able to do, and which it had comparative advantage in. This could ensure efficiency and effectiveness, besides guaranteeing sustainability.

The key government's role in the new arrangement therefore became policy making, and acting as a facilitator of development by assigning roles and responsibilities, keeping order (peace) and security, as well as setting key rules and guidelines (policies) for various sectors. In a number of cases, however, capacity building was necessary to accommodate the changes. Thus some resources were set aside for restructuring of programs to make them compatible with new arrangements. This has made individual partners, government included, more able to play a more positive role, including being more accountable to their partners.

Thus the idea of comparative advantage has made it possible to have positive results trickling down to primary stakeholders, with the government also being appreciated as a facilitator and not a service provider. The questions currently are; are there effective policies to guide development amidst formation of regional economic blocks? And: How do the government policies cope with the conditionalties from other development partners, notably donors? This, however, brings in an element of accountability into focus.

BIBLIOGRAPHY

a. Bolger J (2000). Capacity development: Why, what and how, capacity Development. Occasional series Vol.1, May 2000. CIDA policy Branch: Canada

b. Chambers, R (1983) Rural development: Putting the last first. London

c. Clayton, A 1999). Contracts or partnerships: Working through local NGOs in Ghana and Nepal. Water aid: London.

d. Coates S (2003) Community and Management: A WEDC post-graduate module. WEDC, Loughborough University, UK

e. Deverill P, Bibby S, Wedgwood A and Smout I (2002) Designing water supply and sanitation projects to meet demand in rural and peri-urban communities. Book 1: Concept, Principles and Practice.WEDC, Loughborough University, U.K.

f. DFID (1998) Guidance manual on Water Supply and Sanitation Programmes. WEDC for DFID, Loughborough University, UK.

g. Gosling L and Edwards M.(1995) Toolkits: A practical guide to assessment, Monitoring, Review and Evaluation. Development Manual 5. Save the children, London.

h. Jones D (2001). Conceiving and managing partnerships: A guiding framework, partner notes series. BPN London.

i. Mulwa FW and Nguluu SN (2003) Participatory Monitoring and Evaluation: A strategy for Organizational Strengthening. edn 2. Zapf Chancery, Eldoret, and PREMESE Olivex Publishers, Nairobi, Kenya.

j. TICH (2003). TICH Partnership reports. Tropical institute of community health and development (TICH) in Africa, Kisumu, Kenya.

SITUATION ANALYSIS AND BASELINE SURVEYS
INTRODUCTION:

(i)AIM: To improve water services to people in fringe areas of four small towns in a developing country.

(ii)FACTS AND COMMENTS

(a)   Situation analysis

People at the fringes of the four small towns:

1. Walk long distances (get tired)
2. Use hand pumps or public stand posts (Some cannot use these services due to social prestige, physical handicap, or age)
3. Vendors bringing water to homes of some residents (Never sure of quality & source).

4. Complaint about Long queues (opportunity cost of users' time felt)
5. High price to water vendors (willing to pay for a better service).

(b)   Sustainability of existing facilities:
1. Town council water department managed- (Not sustainable; conflict of interest)
2. Water tariff low- (Unsustainable; low revenue & cannot guarantee improvement)
3. Less revenue to manage O & M costs--- (Expected due to low tariffs; Unsustainable)
4. Unreliable services- (Expected; Limited funds for reliable services)
5. Ministry provides subsidies for O& M and capital costs- (Not sustainable; who pays for it?)
6. Management of water finances poor...(Needs improved accountability, record keeping).
7. Institutional strengthening planned- (Basic necessity for sustainability at stage 1 of planning)
8. Consultancy project team to be engaged- (Community contracting needed).
9. Source of funds for planned project- government plus & donor(s) (Community contribution needed for stronger sense of ownership and community management).

Steps involved in various strategies
(a) NON-FINANCIAL STRATEGY
Initial plans will address:
   i. Ownership- WTP surveys to guarantee O & M support by the users;
  ii. Human resource base- training users in technical and managerial skills;
 iii. Project design & M- Mobilization and training of stakeholders on O & M and create a partnership;
  iv. VLOM- Will prioritize capacity building by forming VLOM and focus groups;
   v. Government capacity- Will enlist government support by empowering the relevant departments to have skills, motivation and resources to support community management;
  vi. Policy environment- will encourage government departments to revise their policies and services to foster community management.

vii. Contract and commitment Local leaders and committee members will sign a contract with the project agreeing to assume responsibilities for O&M.

(b) FINANCIAL STRATEGY

Commercial orientation, demand-responsiveness and participatory water services with improved financial management and a commitment to cost recovery will be our focus.

Finance management

A communication strategy will be designed, and used to conscientise stakeholders about the proposal. Community training on water, hygiene and sanitation will begin immediately to stimulate demand. Preliminary survey will be done to identify project stakeholders, initially by a tentative committee. Then a stakeholder gathering will be called after crop harvest where fundraising and election of committee members will take place. The fundraiser will avail project seed money, which will be used to immediately open a project account by the elected officials. The meeting will decide on subsidy policy.

A finance committee will handle all financial aspects of the project. They will be trained in, and thereafter keep clear simple records of all transactions. They will open and manage the project bank account. The Government and donor will pay the project capital costs, while users will cater for the rest. Records will be available to stakeholders to peruse if they want. They can also be free to get the bank account balance. A simple financial report will be read to and discussed by all members in an annual gathering where new officials are elected. The same will be published in daily papers in at least three languages.

TARIFF SELECTION

Our tariff policy will follow the CAFES principles (Deverill et al, 2002; Coates, 2003)). Tariff will be used to stimulate demand for a higher level of both service and water use. Whereas multi-purpose water use will be encouraged to boost tariff, luxurious consumption will be charged higher, while special groups like the elderly will benefit from subsidy. Special groups will get water using their own means from stand posts or boreholes without any direct payment. Their subsidies will be recovered from other water users. The disadvantaged groups will each be given a monthly waterproof plastic card produced any time they go for water. They will be entitled to a maximum amount of water (quantified as pail-full or jerry can equivalents) every month, and the cumulative amount used recorded in the card whenever used. Non-card holders will pay a fixed rate per litre of water they use right at the water point.

Increasing block tariff will be used for residential yard taps. A Lifeline tariff (LFLT) will be 120 L/HH/day and 60 L/HH/day for lower and higher class residential estates respectively. For industrial and agricultural premises, a tariff 100% of the residential stand post rates will apply, with no LFLT.

The stand post and borehole monies will be collected by the management committee. This will be taken twice a day for banking by another member of the committee who will rotate round five water points on weekdays. On the weekend, the money will be kept in a safe and kept at the nearest police post awaiting banking the following weekday by the same officer. Water meter readings will be taken on the last week of each month. The finance office will prepare bills within 2 days of the readings. Payments will be by cash or cheques at the project office, or project account whose number will always be printed in user receipts and bills. A receipt will be issued for any payment.

CHOOSING / AGREEING ON PREFERRED WATER SERVICE OPTIONS.

To reach collective decision on water points' location and prioritization, consensus building (CB) will be necessary. Focus groups will be used in the CB process. The following are the basic Principles of consensus building (CB):

- Participants to avoid arguing for their own rankings; they need to present their positions clearly, but consider group reactions in subsequent presentations of the same point;
- Avoid a win-lose stalemate- all should view themselves as winners;
- Avoid changing minds only in order to avoid conflict and to reach agreement with harmony;
- Withstand pressures to yield which have no logical foundation;
- Avoid conflict-reducing techniques and treat differences in opinion as indication of incomplete sharing of relevant information on someone's part and press for additional sharing.
- View differences of opinion as helpful and natural rather than a hindrance in decision making;

- View initial agreement as suspect. Explore causes of agreement; ensure that people arrive at similar solutions for either the same basic reasons or for the complementary reasons;
- Avoid subtle forms of influence and decision modifications e.g. when a dissenting member finally agrees, don't feel that he must be rewarded by having his own way on some later point;
- Be willing to entertain the possibility that your group can excel at its task;

Below are some issues likely to arise at CB sessions.

Service option	Drilled borehole and hand pump	Piped supply to water/water Kiosk	Piped supply to yard tap
Level of Service	Lowest	Low	High
Time saving	Lowest	Low	High
Capital cost (Sh / capita)	10-15,000	20-30,000	100-200,000
O & M costs (Sh capita / yr)	5-700	3-5,000	5-10,000
Required level of financial management	High	Higher	Highest
Distance	Long ($\leq$ 250m)	Long	Short
Technical feasibility	Towns A and B	All towns	All Towns
Privacy level	None	None	Excellent
Non-operational time	Longest	Shorter	Shortest
Need to import Spare parts	Yes (pump)	No	No
Tragedy of the commons syndrome	Yes	Less (Manned)	None
Chances of Sustainability	Low	Moderate	High
Chances of child abuse[2]	Very high	High	Low
Payment convenience for Informal sector workers	Yes	Yes	No
Salaried workers	No	No	Yes (Mostly)
Operational convenience	No	No	Very
Operational difficulties[4]	Yes	Yes	No
Tariff collection	Difficult	Easier	Easiest[3]
Intimidation from vendors	V.High	High	None
Supports vendors	High	Moderate	None[5]
Chances of abuse of parts	High	Less	No
Need for subsidy	Yes	No	No
Chance for socializing and marriage deals	High	High	Low
Construction costs	Low and flexible	Moderate	Very high
Construction	Easy	Difficult	Difficult
Convenience to working people	No	No	Yes
Need for Manning	Very high	Very high	None
Popularity	Needs CVM survey	Needs CMV	Needs CMV
Need for coping strategies	Yes (Very high)	High-moderate	No (A little)
Long term cost to user	Highest	Higher	High
Ability to pay tariff	Highest	High-moderate	Moderate
Likely WTP	High	High-moderate	Moderate

1: some men may pump the water to some users for favors.
2: Some children likely to be used by adults to keep positions in queue.
3: Monthly bill; readings taken from water meters.
4: Operational difficulties relate to times of operation; need flexible operational hours;
5: Yard taps likely to render vendors jobless, leading to illicit behaviours to get income.

Conclusion: Towns A and B need to have all the three options installed at points most convenient to prospective users. Locations and numbers of water points should be decided by all stakeholders using maps, CB etc. Yard taps can be installed immediately for the customers WTP in all towns, then. Focus groups used to plan the project's other aspects.

OUTLINES FOR PLANNING AND MONITORING
(a) PLANNING
This table outlines my project planning approach.

Position of project	Project Approval	Planning by mapping, PMR, livelihood analysis, Venn diagrams, focus group discussion, SWOT, Stakeholder analysis, problem trees, objective trees, sanitation ladders and attribute ranking			Implementation	O & M- and Evaluation
Activity (In order of priority)	Communication strategy & education	Training of community members & other stakeholders	Initial draft detailed proposal	Formation of management focus groups	Final detailed project proposal	Use, Pay, Evaluate
Importance	Publicity	Empowerment		Publicity	A CONTRACT	Test of sustainability
	Awareness	Capacity building		-Visits by groups to all stakeholders		
	Social marketing for sanitation programmes	Organizational strengthening		Coordinate stakeholder activities		
Source of funds	Government & donor	All stakeholders	All		All	Community
oles Government Implementing agency Donor Community Politician	Initial funding Assigns responsibilities, Quick WTP assessment Initial funding/ WTP assessment Mobilization Mobilization	All participating in writing different chapters in mixed groups	All Making planning tools And Establishing financial strategy, Motivation, and Technical assistance	-All participating in writing different chapters of proposal as assigned in a participatory system	Use, pay, O & M & consumption Costs	
All stakeholders	Participatory Baseline survey/ feasibility study Including WTP survey using revealed preference Techniques Resource persons in training sessions,	Detailed survey/ feasibility (CVM) -Funding/ community contributions going on -Social marketing Training	Implementing (Each with a specific role) -Funding -Monitoring	-Evaluation -Social marketing -Support		

	Training facilitation, -Funding / collection of funds Formation of a tentative management committee, -Social marketing -Stimulating demand -Tentative budgeting -Monitoring and mobilization	-Budgeting Monitoring Confirmation of management committees			

Community Mapping

Mapping: Is an effort by community to present their before and after perceptions about the shape of their residential area. It is useful when identifying service points (DFID, 1998; Gosling and Edward, 1995; Mulwa and Nguluu, 2003)).

Preference matrix ranking (PMR) (table below).

Properties Option	Constru ct cost	Level of service	Easy to operate	Cheap to O&M	Accident risks	Source of income	Easy to collect tariffs	Income generation prospects
Borehole								
Standpipe								
Yard tap								

PMR will help select the most popular service. It puts caters for characteristics of the variables being ranked, and people make their reasoned choose.

Livelihood analysis: A before and after scenario will help analyze community survival techniques in their diverse situations; e.g. source of income, and expenditure, and be useful guide on tariff setting (Coates, 2003 and Orieko, 1994).

Focus group discussion: Comprises 6-12 selected representatives of the larger community, and will be used in consensus building to select, agree and prioritize the options.

Monitoring: This will be done as a series of mini evaluations. It will cover aspects like interpretation, reporting, and data storage. It will be an ongoing process, to keep track of progress, to improve efficiency, help adjust work plans, focus on inputs, outputs, process outcomes, and work plans. Information sources will be by routine systems, field observation, progress reports and rapid assessments; undertaken by all stakeholders, who report to all stakeholders. I will use monitoring to ensure that Inputs are ready on time; Work plans followed closely; adjustments made, and corrective measures taken where necessary; stakeholders kept informed on time; Constraints and bottlenecks foreseen, timely solutions found; and Resources used effectively and efficiently.

Monitoring tools.
- Group Constitution, Minutes and meetings attendance record;
- Visitors, membership and receipt books
- Activity attendance and Members contribution record

Table: Guidelines for Monitoring of quantitative indicators (Mulwa and Nguluu, 2003)

Item/area	Previous period	1st quarter	2nd quarter	3rd quarter	4th quarter	Total to date	% Attained to date VS planned targets
# Of sub-projects under implementation							
TOTAL							
No. Of sub-projects completed							
Total							
# Of current and potential Beneficiaries							
No of sub-projects with committees							
Gender composition of Sub-project committees Male... Female...							

Comment...............................
Name...................Position...........................Date...............Sign......

Monitoring Qualitative Indicators:
Proxy qualitative indicators will be identified by focus groups, and community gatherings on special occasions to describe, narrate and present (as role play or case studies, or any verbal or written) analytical interpretation of events and trends. Group monitoring will be done using the 7-element cycle outlined below:

A   Change of attitude;
B   Quality of peoples organization;
C   Organizational capacity;
D   Strengthening of peoples economy
E   Improved family standard of living;
F   Networking and linkages; and
G   Participatory reviews

Each item will be scored as three for good; two for fair; one for poor, and as many stakeholders as possible will give their responses to these items Mulwa and Nguluu, 2003).

4 References

o Coates S (2003) Community and Management: A WEDC post-graduate module. WEDC, Loughborough University, UK
o Deverill P, Bibby S, Wedgwood A and Smout I (2002) Designing water supply and sanitation projects to meet demand in rural and peri-urban communities. Book 1: Concept, Principles and Practice.WEDC, Loughborough University, U.K.
o DFID (1998) Guidance manual on Water Supply and Sanitation Programmes. WEDC for DFID, Loughborough University, UK.
o Gosling L and Edwards M.(1995) Toolkits: A practical guide to assessment, Monitoring, Review and Evaluation. Development Manual 5. Save the children, London.
o Mulwa FW and Nguluu SN (2003) Participatory Monitoring and Evaluation: A strategy for Organizational Strengthening. edn 2. Zapf Chancery, Eldoret, and PREMESE Olivex Publishers, Nairobi, Kenya.
o Orieko PC (1994) Community Development: Its concepts and practice with emphasis on Africa. Gideon S Were press, Nairobi
o Unknown author: Guidelines to consensus building.

SECTION 2: ENVIRONMENTAL ASSESSMENT
Introduction

Basic checklist used to compile the description of the environmental setting
This checklist lists some factors, which should be considered in describing the environment. This description of the environmental setting is a record of conditions prior to implementation of the proposed project. It is primarily a benchmark against which to measure environmental changes and to assess impacts.

1. BASIC LAND CONDITIONS
a. Geological Conditions

Major land formations (valleys, rivers)
Geologic structures (sub-strata, etc.)
Geologic resources (minerals, oil, etc.)
Seismic hazards (faults, liquefaction, tidal wave etc.)
Slope stability and landslide potential

b. Soil Conditions
➤ Soil conservation service, classification
➤ Hazard potential (erosion, subsidence or expansiveness)
➤ Natural drainage rate
➤ Sub-soil permeability
➤ Run-off rate
➤ Effective depth (inches)
➤ Inherent fertility
➤ Suitability for method of sewage disposal

c. Archaeological value of site

2. BIOTIC COMMUNITY CONDITIONS
a. Plant
➤ General type and dominant species
➤ Densities and distributions
➤ Animal habitat value
➤ Historically important specimen
➤ Watershed value
➤ Man-introduced species
➤ Endangered species (location, distribution and conditions)
➤ Fire potential (chaparral, grass, etc.)

- ➤ Timber value
- ➤ Specimen of scientific or aesthetic interest

b. Animal
- ➤ General types/dominant species (mammal, fish, fowl, etc.)
- ➤ Densities and distribution
- ➤ Habitat (general)
- ➤ Migratory species
- ➤ Game species
- ➤ Man-introduced species (exotic species)
- ➤ Endangered species
- ➤ Commercially valued species

3. WATERSHED CONDITIONS
- ➤ Water quality (ground water and surface water)
- ➤ Source of public or private water supply on-site
- ➤ Watershed importance (on-site and surrounding area)
- ➤ Flood plain importance (on-site and surrounding area)
- ➤ Water run-off rate
- ➤ Streamside conditions (habitat conditions and stream flow rate)
- ➤ Location of wells, springs
- ➤ Marshlands, lakes, ocean frontage importance

4. AIRSHED CONDITIONS
- ➤ General climatic type
- ➤ Air quality
- ➤ Airshed Importance
- ➤ Wind hazard area (min/max speeds)
- ➤ Odour levels
- ➤ Noise levels
- ➤ Rainfall (average)
- ➤ Temperature (average highs and lows)
- ➤ Prevailing winds (direction and intensity)
- ➤ Fog conditions (hazard potential)

THE ENVIRONMENTAL ASSESSMENT PRACTICE IN KENYA
Introduction:
The EIA is a critical examination of the effects of a project on the environment. An EIA identifies both negative and positive impacts of any development activity or project, how it affects people, their property and the environment. EIA also identifies measures to mitigate the negative impacts while maximizing positive ones. EIA is basically a preventive

process, which seeks to minimize adverse impacts on the environment and reduces risks (GoK (1999 and GoK (1994). If a proper EIA is conducted, then the safety of the environment can be properly managed at all stages of a project- planning, design, construction, operation, monitoring and evaluation as well as decommissioning NEMA (2004, 2006, 2005, 2003a, 2003b)

Environmental Impact Assessment Guidelines

The environmental impact assessment guidelines require that environmental impact assessment be conducted in accordance with the issues and general guidelines spelt out in the second and third schedules of the regulations NEMA (2004, 2006, 2005, 2003a, 2003b)

Goal and objectives of EIA:

According to NEMA (2004, 2006, 2005, 2003a and 2003b), and GoK (1999 and Gok, 1994), the goal of EIA is to ensure that decisions on proposed projects and activities are environmentally sustainable.

The objectives of EIA are:

1. To identify impacts of a project on the environment;
2. To predict the likely changes on the environment as a result of the development;
3. To evaluate the impacts of the various alternatives on the project (including a no project alternative);
4. To propose mitigation measures for the significant negative impacts of the project on the environment
5. To generate baseline data for monitoring and evaluating impacts, including mitigation measures during the project cycle;
6. To highlight environmental issues with a view to guiding policy makers, planners, stakeholders and government agencies to make environmentally and economically sustainable decisions.

Which projects require EIA?

The projects to be subjected to EIA are specified in the second schedule of EMCA 1999 (GoK (1999) and GoK (1994). This includes coverage of the issues on schedule 2 (ecological, social, landscape, land use and water considerations) and general guidelines on schedule 3 (impacts and their sources, project details, national legislation, mitigation measures, a management plan and environmental auditing schedules and procedures)(NEMA (2004, 2006, 2005, 2003a and 2003b),The full list is as shown below:

i.   General: (i) an activity out of character with its surroundings; any structure of a scale not in keeping with its surrounding; Major changes in land use.
ii. Urban development
iii. Transport
iv. Dams, rivers, water resources;
v.   Aerial spraying;
vi. Mining, including quarrying and open cast extraction;
vii. Forestry related activities;
viii. Processing and manufacturing industries;
ix. Electrical infrastructure
x.   Management of hydrocarbons;
xi. Waste disposal;
xii. Natural conservation areas;
xiii. Nuclear reactors
xiv. And major developments in biotechnology

When to conduct EIA:

EIA is considered part of the project development process, and is thus done at the initial stages of project development. . It is a decision making tool and should guide whether a project should be implemented, abandoned or modified prior to implementation. Issues considered in the EIA as per the EMCA 1999(NEMA (2004, 2006, 2005, 2003a and 2003b) include the following:

• Ecological considerations, including biodiversity, sustainable use, and ecosystem maintenance;

• Social considerations, including: economic impacts; social cohesion or disruption; effects on human health; immigration or emigration; communication; and effects of culture or objects of cultural value;

• Landscape issues, including views opened up or closed, visual impacts; compatibility with surrounding area; and amenity opened up or closed;

• Land use, including: effects on current land uses and land use potentials in the project area;

• Effects of proposal on surrounding land uses and land use potentials; and possibility of multiple uses.

• Water: water resources (quality and quantity of sources; and drainage patterns or systems.

The EIA process:

Development and submission of a project report for projects or activities which are not likely to have significant environmental impacts or those for which EIA study is required (GoK (1999 and GoK 1994). Then the EIA process, if done, is as follows (NEMA (2004, 2006, 2005, 2003a and 2003b).

➢ Scoping and drawing up terms of reference (TOR) for the study for approval by the authority;
➢ Gathering baseline information through investigation, research and subsequent submission of EIA report to the authority. At submission, ten copies of the report, a copy of the report in Compact Disc, and a banker's cheque for 0.1% of the project cost (or Kshs 10,000 whichever is higher) payable to NEMA. The reports are thereafter distributed to NEMA review experts for peer review, normally within 21 days of project submission.
➢ Review of EIA study report by the authority and NEMA recommended lead agencies.
➢ Decision on the EIA study report, which may include non-conditional approval, conditional approval, or rejection.
➢ If approved, then the significant impacts and their proposed mitigation measures are prepared by NEMA and sent to the proponent to advertise in the media for further stakeholder input. A maximum of 90 days are allowed to pass during which these views are submitted to NEMA. In the absence of a negative concern from stakeholders within the 90 days, an EIA certificate is issued to the proponent.
➢ Implementation of the project / OR Appeals: The latter applies if the project is not approved.
➢ Monitoring the project (Based on the indicators identified at the EIA)
➢ Auditing the project (Environmental audit).

Who administers EIA?
NEMA is mandated by the EMCA number 8 of 1999 to administer the EIA (GoK (1999 and GoK, 1994).

Who conducts EIA?
Individual experts or firms of experts registered by NEMA are the only ones mandated top do EIA / EA studies. A register of EIA experts is available in the authority's headquarters, district provincial offices and can be accessed upon payment

of a fee of Ksh 200 (USD 2.8) (NEMA (2004, 2006, 2005, 2003a and 2003b).

Public participation in EIA / EA studies:
The law requires that during the EIA process, a proponent shall, in consultation with the authority, seek the views of persons who may be affected by the project or activity through posters, newspapers, and radio; hold at least three public meetings with the affected parties and communities. The public participates either by submitting written proposals or making oral comments. Such comments are considered in reviewing the EIA study report (NEMA (2004, 2006, 2005, 2003a and 2003b).

Who pays for EIA?
The project proponent pays for the entire EIA process. It identifies and recruits and pays the consultant, who must be a NEMA registered and licensed expert. Thereafter, the proponent, through its consultant, pays 0.1% of the project cost to NEMA when the reports are being submitted to NEMA (NEMA (2004, 2006, 2005, 2003a and 2003b),.

EIA Fee payable:
➤ Lead expert pays a registration fee of Kshs 3000 (9000 if non citizen), and an annual practicing license of Ksh 5,000 (15000 if non citizen);
➤ Associate expert pays a registration fee of Kshs 2000 (6000 if non citizen), and a n annual practicing license of Ksh 3,000 (9000 if non citizen);
➤ A firm of experts pays a registration of Ksh 5,000, and an annual practicing fee of 20,000;
➤ An EIA license fee of 0.1% of total project cost payable to NEMA by proponent (nothing is paid by proponent for EA);
➤ EIA license surrender, transfer or validation fee of Ksh 5,000

Compliance to EA:/EIA (NEMA (2004, 2006, 2005, 2003a and 2003b),
Generally, the following apply as far as compliance is concerned:
1. A proponent shall not implement a project likely to have a negative environmental impact, or for which an EIA is required by the EMCA 1999 or regulations issued under it

unless an EIA has been concluded and approved according to law

2.	No licensing authority under any law in force in Kenya shall issue a trading, commercial or development permit for any project for which an EIA is required or for a project / activity likely to have a cumulative significant negative environmental impact unless the applicant produces an EIA license issued by the authority.

Current limitations of the EIA as an Environmental legislation and process

The EIA legislation is too specific, and recommends an EIA. EIA, however, is project based, and has serious limitations shown below. Instead, a strategic environmental assessment approach could be taken. Strategic Environmental Assessment (SEA) is an improvement of the formal Environmental Impact Assessment (EIA) involving the analysis of policies, plans and programs. The EIA only deals with project-based impacts. Thus EIA is short-lived, and addresses only issues directly relevant to the project, within its cycle from conception, up to evaluation. Secondly, the EIA mostly stops immediately after project approval by the relevant authority i.e., at decision-making stage. SEA addresses cumulative (additive and synergistic) effects of plans, policies and programs in a wider framework that would also cover project-based assessments. SEA sets the legal, institutional, economic, social and environmental framework for broader operation of EIAs ((Lee and George (2000), Dijkstra (2003) and Canter (1977)).

Due to these limitations of the EIA, SEA strives to fill some of the gaps. The SEA approach deals with impacts in a wider scale, and covers more time; it is a continuous process as the assessments are part of an institution's policies, plans and programs. Generally, projects, programs, plans and policies have varied scope, with the project being the lowest and policy is the highest.

A policy may cover a whole country or region, and therefore tends to have a cumulative (i.e., temporal and spatial) consideration of impacts. Project-based EIAs, on the other hand, address narrower site-based or local impacts (canter, 1996). Since the policies are wider in scope, they tend to cut across many projects, plans and programs. On the other hand, a program covers a number of projects, while a plan covers a number of programs. Therefore program, plan and policy-based environmental assessments are relevant to and have the capacity to cover more projects than the project-based assessments (Dijkstra T (2003).

When SEA is embedded in a country's legislature, it has the capacity to reduce costs associated with project assessment and costs, since some impacts assessed from earlier projects within that wider framework can be applied to a number of projects; this caters for a wider networking. Similarly, through SEA, broad guidelines can be institutionalized to ascertain minimum standards of the environmental assessment. SEA, being wider in scope, is likely to cover a wider area, and can therefore have more capacity to handle impact assessments, which gives it more credibility (Dijkstra, 2003). SEA also facilitates decision-making (especially once the analytical environmental assessment is incorporated), thereby reducing project delays and harmonizes the entire project cycle, so that the direction of any project is certain depending on its nature, location and scope.

EA can give earlier guidelines, going by the experience from related or similar projects, on what EIA steps can be skipped, and which are mandatory. This saves time and resources, eases the screening and scoping stages, and makes it possible for the other stages of the EIA to be applied in time from the project conception through to the end. This is likely to provide some social and economic gains to the community, and improve the general standard of living since the reduced cost of project implementation may benefit the users of the product or service from that project.

Lastly, due to clear inscription of environmental assessment guidelines in national and regional records, wider participation in the process is likely, as the basis of each project will already be clear to a section of the stakeholders. This can act as a conflict-resolution strategy. However, these assessment guidelines can be availed in local languages by print and alternative media to facilitate wider participation, since in the developing countries, the EIA is either new, or non-existent (Lee and George, 2000). The SEA can therefore be a reasonable substitute for EIAs, a role they can perform even more effectively. In Botswana, for instance, EIA is not yet mandatory (GoB, 2003). Yet the numbers of new projects being developed are so many that unless their long-term repercussions are addressed early through alternative means such as SEA, there projects can only be time bombs.

Challenges in the EIA administration:
1. Lack of compliance by Government agencies:
   i. The EIA process is nobly managed by a government parastatal, the authority. Whereas it is mandatory for all relevant projects to undergo EIA, the government institutions have never complied. This has been attributed to the following factors, as per personal observation and communication) (NEMA (2004, 2006, 2005, 2003a and 2003b),:
   ii. Do not consider NEMA as superior to them to warrant NEMA involvement in their affairs- in an activity akin to supervision;
   iii. NEMA has no machinery to ensure compliance by government agencies;
   iv. The agencies are aware that the government has no capacity to take disciplinary actions against a department of its own, e.g. will not close them down. Fear of closure by government for non-compliance is a major facilitator of compliance by non-government bodies.
   v. Rate of non-compliance even by non-government bodies is still high, necessitation a close follow up. Thus the numbers overwhelm NEMA;
   vi. Political interference even in cases where due EIA process has been followed, thus intimidating the NEMA management and making it feel less likely to intervene in government projects;
   vii. Negative publicity by senior government ministers about NEMA as scaring away investors.

Up to April 2006, 1080 projects had undergone an environmental impact assessment. This was then estimated to comprise about 75% of projects qualified for EIA.

Corruption by proponent
Many proponents try to bribe the EIA experts, sometimes successfully, with a view to excluding some adverse impacts from the study report. This is done by reducing the

thoroughness of public participation through various mechanisms, including:
Selecting for the consultant his / her audience during the EIA process. This includes: who to talk to, interview, invite in public meetings, or include in group discussions. This has the capacity to exclude some concerns, or even some positive impacts of the project.

Possible Solution: The consultant should stay beyond reproach, follow due process and conduct the work professionally.

Corruption by consultant:
While trying to make the proponent to pay more, some consultants arm twist the proponents by scaring them that the project has too much significant impact, and stand no chance to be approved. They go ahead to tell them that if they paid extra, they would talk to NEMA so that the project goes through. The victims are largely illiterate and the very busy business people, who understand least of the existing legislation.

Ignorance of the public about their roles in EIA and EA:
Most of the time, when the stakeholder consultation and public participation process is on, many fear to give negative comments fearing that they would be incriminated.

Solution: Many consultants have allayed these fears by first educating the publics the importance of openness of all parties in the process. Holding the meetings on neutral grounds, as well as building the confidence of the publics, helps make them open.

Project gaps and dishonesty:
In some cases, the proponent has deliberately not given full, clear information on the project to the consultant, who then faces the public with incorrect or incomplete information. Based on this, the public cannot give proper feedback. In some cases, the proponent has not quite confirmed some details of the project. However, amidst consultant scramble for the projects, the proponent only gives what they have. Further details of the project sometimes, even core aspects, come when it is too late.

One needs to ensure public participation as follows:

(a) Draft a project proposal: This should give all salient features of the proposed project in simple language, in local language as well. This will include maps and other visual aids (Lee and George (2000), Dijkstra (2003), and Canter (1977).

(b) Form an acting project team: This will incorporate all obvious stakeholder representatives.

(c) Identify all groups of stakeholders: Done by the acting project team. A combination of methods such as (i) those who volunteer themselves; (ii) those identified by the project team; (iii) those identified by other third parties e.g. key groups in the community; etc will be used.

(d) Broaden the project team by incorporating newly identified stakeholders. Let each group of stakeholders be free to have any one of their own to represent them, even if it means rotational representation. But let it be known to the group about the importance of continuity in representation, as any break in communication and information flow may disorient the team.

(e) Make a stakeholder consultation and CPP plan (done by the newly expanded project team / now called working project team (WPT)). This will involve a combination of interviews; questionnaires; rapid appraisal technique; organized sessions (with individuals, groups and entire community, i.e. the public), etc (Lee and George, 2000).

(f) Make the program available to members in public places and strategic popular sites e.g. an area designated as an open door (a site where all information about the programme can be accessed by any interested party)

(g) The program contents will be communicated to the public using different languages, fora and places. Verbal (door- to door), print (local newspapers, posters) and electronic media (Radio, Television, Email, internet website) will be used to communicate the message so that the schedules of CPP are properly publicized. Also, there will be community notice boards specifically for the project updates and information meant to make the community come forward to participate.

(h) Short presentations like concerts, role plays etc in the electronic media (e.g. radio, television etc), will be made to attract the attention of the public towards the project, and inform them about its plan, and how it may affect them- both positively and negatively (Canter (1977); Dijkstra (2003). This is meant to attract attention, so that the public who are not aware that they may be affected may also come forward.

(i) Once per day, during and for one week preceding any gathering, vehicles mounted with loudspeakers will be used to advertise the schedules.

(j) Once the gathering is in place, participatory methods (e.g. use of small groups, role plays, community mapping etc) will be used to gather the concerns, fears, issues, and other information relevant to the project- now and in the future (Coates, 2003). Each group will be asked to select their secretary, who will put down all their points, and later one member of each group will be asked to present their report verbally to the audience. Group discussions will ensure gender sensitivity. Information from the Small groups will be augmented with extra information put in writing by individuals who may have points they may not want to raise in public (either due to their sensitivity, or due to personality /disposition of the individuals). Once verbal presentations are over, the written group reports will be submitted to the secretariat of the project (whose members will also be in different groups, rotating, with one of them as a facilitator)

(k) Report will be compiled using the small reports, and verbal presentations. This will be summarized in point form and its contents made public within a day. The report will be in as many languages as is feasible to facilitate sustainability and financial constraints. This repost will be available at the secretariat, and some copies will be availed free to the public.

REFERENCES:
1. NEMA (2004) Personal communication with a provincial NEMA coordinator, Kisumu, Kenya
2. NEMA (2006) Personal communication with a provincial NEMA coordinator, Kakamega, Kenya
3. NEMA (2005) Personal communication with a national EIA/EA coordinator, Nairobi, Kenya
4. NEMA (2003a) Environmental Impact Assessment (EIA) brochure, Nairobi, Kenya
5. NEMA (2003b) Environmental Audit (EA) brochure, Nairobi, Kenya
6. GoK (1999).The Environment Management and Coordination act (EMCA, 1999). Nairobi
7. GoK (1994) National Environment Management Plan (NEAP), Nairobi.
8. Canter, L W (1996). Environmental Impact Assessment. McGraw Hill International Editions: Civil Engineering Series. 2nd edition. Singapore. ISBN 0-07-114103-0
9. Dijkstra T (2003) Environmental Assessment. A WEDC Postgraduate module. WEDC Loughborough University, Leicestershire.

10. Lee N and George C (2000). Environmental Assessment in Developing and Transitional Countries. John Wiley and Sons, Manchester. ISBN 0-471-98557-0

THE ENVIRONMENTAL LEGISLATION AND ENFORCEMENT IN KENYA

General objectives:
To discuss the environmental practice in Kenya

Specific objectives:
To discuss the environmental legislative framework in Kenya;
To discuss enforcement institutions and mechanisms in Kenya
To discuss the effectiveness of the current environmental legislation in Kenya.

Methods:
Literature review and analysis are used, coupled with case studies.

The Environmental Management and Co-ordination Act (EMCA)

The Environmental Management and Co-ordination Act No. 8 of 1999 is an Act of Parliament that provides for the establishment of an appropriate legal and institutional framework for the management of the environment. As earlier provided, prior to its enactment in 1999, there was no framework environmental legislation. Kenya's approach to environmental legislation and administration was highly sectoral and legislation with environmental management components had been formulated largely in line with natural resource sectors as aforementioned.

EMCA was developed as a framework law, and this is due to the fact that the Act is thus far, the only single piece of legislation that contains to date the most comprehensive system of environmental management in Kenya. The Act provides for the establishment of an appropriate legal and institutional framework for the management of the environment in Kenya and for matters connected therewith and incidental hereto. The Act is based on the recognition that improved legal and administrative co-ordination of the diverse sectoral initiatives is necessary in order to improve national capacity for the management of the environment, and accepts the fundamental principle that the environment constitutes the foundation of our national, economic, social, cultural and spiritual advancement.

Section 3 of the Act enunciates the General Principles that will guide the implementation of the Act. Every person in Kenya is entitled to a clean and healthy environment and has the duty to safeguard and enhance the environment. It is worth noting that the entitlement to a clean and healthy environment carries a correlative duty. Hence, there is not only the entitlement to a clean and healthy environment, but also the duty to ensure that the environment is not degraded in order to facilitate one's own as well as other persons' enjoyment of the environment.

The Act established a number of institutions for the management of the environment in Kenya. At the apex is of course the parent Ministry. Below this is the National Environment Council established under Section 4 of the Act as the body responsible for policy formulation and directions for purposed of the Act, as well as the setting of national goals and objectives and the determination of policies and priorities for the protection of the environment. The Council is chaired by the Minister responsible for environmental matters and its composition is drawn from all public, private and non-governmental sectors of the country as is provided for within Section 4.

NEMA is managed by a Board of Management which comprises fourteen members inclusive of the Chairman, the Permanent Secretary of the Ministry of Environment and Natural Resources, seven members each representing the Provinces of Kenya, three Directors of the Authority and the Secretary of the Board and the Director General manage NEMA. Within NEMA we have the Director General's Office (CEO) and various departments that perform the day-to-day running of NEMA. EMCA also establishes several statutory committees namely Standards and Enforcement Review Committee, the National Environment Action Plan Committee and the Environmental Impact Assessment – Technical Advisory Committee. In addition, there is also the Provincial and District Environment Committees. EMCA also establishes the National Environment Trust Fund and the National Environment Restoration Fund. The main object of this Fund is to facilitate research intended to further the requirements of environmental management, capacity building, environmental awards, environmental publications, and scholarships and grants. In addition, the object of the Restoration Fund is to act as supplementary insurance for the mitigation of environmental degradation. It will be used in cases where the perpetrator of the damage is not identifiable or exceptional circumstances force the Authority to intervene in the control or mitigation of environmental degradation.

All these Committees and feed into each other and connected in one way or another. However, EMCA has provided for the establishment of 2 independent entities namely the Public Complaints Committee and the National Environment Tribunal. The Committee whose functions include the investigation of allegations or complaints against any person or against NEMA in relation to the condition of the environment in Kenya or any suspected cases of environmental degradation. The Tribunal's primary functions are to hear disputes of a technical nature on the administration of the Act as well as appeals against the administrative decision taken by NEMA and other organs responsible for enforcement of the Act and regulations or requirements thereunder. Such appeals may be launched by any person aggrieved by, for instance the refusal to grant a licence under the Act, the imposition of an environmental restoration or improvement order or the quantum of fees he is required to pay under the Act.

The National Environment Management Authority (NEMA) is established under Section 7 of the Act. NEMA is established as a body corporate with perpetual succession, capable of suing and being sued, holding and disposing of property, borrowing money, an entering into contracts in its corporate name.

ENVIRONMENT^AL PROTECTION ENFORCEMENT IN KENYA
THE NATIONAL ENVIRONMENT MANAGEMENT AUTHORITY (NEMA)
Object and Purpose of NEMA
NEMA is the institution with the legal authority to exercise general supervision and co-ordination over all matters relating to the environment, and is the principal instrument of

the Government charged with the implementation of all policies relating to the environment.

Functions of NEMA
NEMA's functions, which determine its scope of activities, are more particularly set out in Section 9 (2) of the Act. They include co-ordination the various environmental management activities being undertaken by the lead agencies and promoting the integration of environmental consideration into development policies, plans, programs and projects; undertake in co-operation with relevant lead agencies programs intended to enhance environmental education and public awareness about the need for sound environmental management, publish and disseminate manuals, codes or guidelines relating to environmental management, prepare and issue an annual report on the state of the environment in Kenya and in this regard may directed any lead agency to prepare and submit to it a report on the state of the sector of the environment under the administration of the lead agency establishing and reviewing, in consultation with relevant lead agencies, land use guidelines; advising the Government on legislative and other measures for the management of the environment or the implementation of relevant international conventions, treaties and agreements in the field of environment; advising the Government on regional and international environmental conventions, treaties and agreements to which Kenya is a party, mobilizing and monitoring the use of financial and human resources for environmental management; and rendering advice and technical support where possible to entities engaged in national resources management and environmental protection so as to enable them to carry out their responsibilities satisfactorily.

As such it is very clear that part of NEMA's role is to ensure that the public receives adequate information on aspects of environmental management in Kenya.

PROVINCIAL AND DISTRICT ENVIRONMENT COMMITTEES
The Act also establishes as per section 29, the Provincial and District Environment Committees of NEMA. These Committees are created for purposes of discussion and decision-making on matters relating to the proper management of the environment within the respective

province or district. The composition of these committees is such that local communities who are closely connected to the resources are empowered to have a say in their management. Among the members of the PEC are two representatives of farmers or pastoralists within the province, 2 representatives of the business community within the province, 2 representatives of NGO engaged in environmental management programs within the province. With respect to the DEC, membership includes 4 representatives of farmers, women, youth and pastoralists within the district, 2 representatives of NGOs engaged in environmental management programs in the districts and 2 representatives of community based organizations engaged in environmental programmes in the district. This elaborate constitution of the committees not only enhances the roles of individuals and community-based groups, but also facilitates public participation in the decision-making process and thereby boosts environmental governance.

THE NATIONAL ENVIRONMENT ACTION PLAN
 Public participation in environmental management is further enhanced by the requirements under Sections 39 and 40 of the Act for every Provincial and District Environment Committee to prepare every five years, a provincial and district environment action plan for the respective province and district for which the committee is appointed.

The Environment Action Plans are required to contain inter alia an analysis of the natural resources for each province or district with an indication as to any pattern of change in their distribution and quantity over time, a recommendation of appropriate legal and fiscal incentives that may be used to encourage the business community to incorporate environmental requirements into their planning and operational processes, proposed guidelines for the integration of standards of environmental protection into development planning and management and identification and recommendation of policy and legislative approaches for preventing, controlling or mitigating specific as well as general adverse impacts on the environment. The makeup of the Committee is such that the public, private and non-governmental sector are represented and the views of these bodies are taken into consideration in the development of the Action Plan.

PROTECTION AND CONSERVATION OF THE ENVIRONMENT

Part V of EMCA provides legal tools for sustainable management of the environment. It covers the protection and management of wetlands, hilly and mountainous areas, forest, environmentally significant areas, the ozone layer and the coastal zone. It further provides for the conservation of energy and biological diversity, access to genetic resources and environmental incentives. This Part of EMCA delegates onto the director general. Various responsibilities to ensure protection and sustainable management of the environment. In addition, the part also gives the Minister in charge of environmental affairs the mandate to give orders, directions or regulations and standards vide gazette notice.

ENVIRONMENTAL IMPACT ASSESSMENT

The importance of public participation in decision-making in environmental matters is further highlighted by the requirement for environmental impact assessment study report under Part VI of the Act. Any person, being a proponent of a project is required to apply for and obtain an E.I.A licence from NEMA before he can finance, commence, proceed with, carry out, execute, or conduct any undertaking specified in the 2nd Schedule of the Act. The EIA study report is published for two successive weeks in the Gazette and in a newspaper circulating in the area or proposed area of the project and the public is given a maximum period of sixty days for inspection of the report and submission of oral or written comments on the same. Any person may extend this period on application. The EIA process, thus, gives individuals and communities a voice in issues that may bear directly on their health and welfare and entitlement to a clean and healthy environment.

ENVIRONMENTAL AUDIT AND MONITORING

Part 7 of the Act (Sections 68-69) gives NEMA the responsibility of carrying out environmental audits of all activities that are likely to have significant effect on the environment. In consultation with lead agencies, the Act also authorises NEMA to carry out environmental monitoring of all environmental phenomena and operations of industry, projects or activities to determine their impacts.

STANDARDS AND ENFORCEMENT COMMITTEE

Another statutory committee is the Standards and Enforcement Review Committee established under Section 70 of the Act and chaired by the Permanent Secretary under the Minister responsible for environmental matters. The functions of the Committee includes advising NEMA on how to establish criteria and procedures for the measurement of water quality, recommending to NEMA minimum water quality standards for all waters of Kenya, analysing and submitting to the Director General conditions for discharge of effluents into the environment, and documenting the analytical methods by which water quality and pollution control standards can be determined and appointing laboratories for the analytical services required.

Other statutory functions of the committee are advising NEMA on how to establish criteria and procedures for the measurement of air quality (Section 78), the issue of regulations and guidelines and the prescription and submission to NEMA of draft standards on pesticides and toxic substances (Section 94), recommending to the Authority standards for emissions of noise and vibration pollution into the environment (Section 101), the establishment of standards for ionising and other radiation (Section 104). In this respect the Act confers on the Standards and Enforcement Committee rulemaking powers. This is important in the light of the fact tat regulations and rules are required to implement the framework provisions of the Act. In practice, the draft regulations and standards are adopted by the Board of Management for ownership and then forwarded to the Minister for promulgation and gazettement. The Committee therefore acts as the technical arm of NEMA in setting these standards.

PUBLIC COMPLAINTS COMMITTEE
This institution is set up independent of the Directorate General of NEMA. It is probably unique to Kenya's institutional framework with respect to the settlement of disputes related to environmental management that is access to justice. Section 31 of the Act establishes the Public Complaints Committee as a Committee of NEMA, headed by a person qualified for appointment as a High Court judge and appointed by the Minister. The Committee whose functions include the investigation of allegations or complaints against any person or against NEMA in relation to the condition of the

environment in Kenya or any suspected cases of environmental degradation. It is a quasi-judicial body with powers to subpoena witnesses and compound offences.

It functions like an environmental ombudsman. Upon investigating and hearing the public on their complaints, the Committee is required to make a report of its finding and recommendation thereon to the National Environment Council which, in turn, directs NEMA on the implementation of the recommendations. The Committee provides a process for legal empowerment of the public as well as a forum to check and control private and public policies and decisions in order to ensure environmental accountability and consequently, sound environmental management. The Committee has been set up in such a way as to allow for easy access by the general public. No elaborate mechanisms as necessary for it to start investigation of any complaint.

NATIONAL ENVIRONMENT TRIBUNAL

The second independent institution established under EMCA is the National Environment Tribunal. This institution is established under Section 125 of the Act and chaired by a person qualified for appointment as a judge of the High Court of Kenya, nominated by the Judicial Service Commission. The Tribunal's primary functions are to hear disputes of a technical nature on the administration of the Act as well as appeals against the administrative decision taken by NEMA and other organs responsible for enforcement of the Act and regulations or requirements thereunder. Such appeals may be launched by any person aggrieved by, for instance the refusal to grant a licence under the Act, the imposition of an environmental restoration or improvement order or the quantum of fees he is required to pay under the Act.

The Act provides that the Tribunal shall not be bound by the rule of evidence applicable in regular judicial proceedings, thereby making the Tribunal's proceedings non-technical in nature and hence accessible to every person. The Tribunal is empowered to make such orders and awards as it may deem just. Any person aggrieved by the order or award of the Tribunal may appeal to the High Court which has powers to confirm, set aside or vary the decision of the Tribunal or to remit the proceedings of the Tribunal for further consideration, or to make such other order as it may deem just. The Tribunal has been established as an open forum easily accessible to all persons in Kenya and does not adhere to the strict judicial processes as is required in the Courts of Law. However, the decisions of the Tribunal are binding to all persons affected.

To facilitate this process, regulations on environmental impact assessment and environmental audits have been established under the Kenya Gazette Supplement No. 56 of 13th June, 2003. Besides, a number of other national policies and legal statutes have been reviewed to enhance environmental sustainability in national development projects across all sectors. Some of the policy and legal provisions are presented in the following sub-sections.

Policies
National Environmental Action Plan (NEAP)
According to the Kenya National environment Action Plan (NEAP, 1994) the Government recognized the negative impacts on ecosystems emanating from industrial, economic and social development programmes that disregarded environmental sustainability. Following on this, establishment of appropriate policies and legal guideline as well as harmonization of the existing ones have been accomplished and/or are in the process of development. Under the NEAP, process Environmental Impact Assessment was introduced and among the key participants identified were the industrialists, business community and local authorities.

National Policy on Water Resources Management and Development
While the National Policy on Water Resources Management and Development (1999) enhances a systematic development of water facilities in all sectors for the promotion of the country's socio-economic progress, it also recognizes the by-products of these processes as water. It, therefore, calls for the development of appropriate sanitation systems to protect people's health and water resources from institutional pollution.

Industrial and business development activities, therefore, should be accompanied by corresponding waste management systems to handle the wastewater and other wastes emanating from there. The same policy requires that such projects should also undergo comprehensive EIAs that will provide suitable measures to be taken to ensure environmental resources and peoples health in the immediate neighbourhood and further downstream are not negatively impacted by any proposed projects. As a follow-up to this, EMCA, 1999 requires annual environmental audits to ensure continuous improvements on the recommendations from the EIAs.

In addition, the policy provides for charging levies on wastewater based on quantity and quality (similar to polluter-pays-principle) of effluent. Further, the policy requires those contaminating water to meet the appropriate cost on remediation, though the necessary mechanisms for the implementation of this principle have not been fully established under the relevant acts. However, the policy provides for establishment of standards to protect water bodies receiving wastewater, a process that is ongoing.

Policy Guidelines on Environment and Development
Among the key objectives of the Policy Paper on Environment and Development (Sessional Paper No. 6 of 1999) are:
▪ To ensure that from the onset, all development policies, programmes and projects take environmental considerations into account.
▪ To ensure that an independent environmental impact assessment EIA) report is prepared for any industrial venture or other development before implementation,
▪ To come up with effluent treatment standards that will conform to acceptable health guidelines.

Under this paper, broad categories of development issues have been covered that require sustainable approach. These issues include the waste management and human settlement sectors. The policy recommends the need for enhanced re-use/recycling of residues including wastewater, use of low non-waste technologies, increased public awareness raising and appreciation of clean environment. It also encourages participation of stakeholders in the management of wastes within their localities. Regarding human settlement, the paper encourages better planning in both rural and urban areas and provision of basic needs such as water drainage and waste disposal facilities among others.

Legal Aspects

Application of national statutes and regulations on environmental conservation suggest that developers have a legal duty and responsibility to discharge wastes of acceptable quality to the receiving environment without compromising public health and safety, and any related biodiversity. This position enhances the importance of an environmental audit for the operations at the Village Market complex to provide a benchmark for its sustainable operation. The key national laws that govern the management of environmental resources in the country have been discussed in the following paragraphs. Note that wherever any of the laws contradict each other, the Environmental Management and Co-ordination Act 1999 prevails.

The Environment Management and Co-ordination Act, 1999

Part II of the Environment Management & Co-ordination Act, 1999 states that every person in Kenya is entitled to a clean and healthy environment and has the duty to safeguard and enhance the environment. In order to ensure that this is achieved part VI, section 58, of the same Act directs that any proponent of a new project should carry out an environmental impact assessment and prepare an appropriate report for submission to the National Environmental Management Authority (NEMA), who in turn issues a license as appropriate. The second schedule of the same Act lists proposed urban development activities as among the facilities that should undergo environmental impact assessments.

Part VIII, section 72, of the Act prohibits discharging or applying poisonous, toxic, noxious or obstructing matter, radioactive or any other pollutants into aquatic environment. Section 73 requires that operators of projects, which discharge effluent or other pollutants, submit to NEMA accurate information about the quantity and quality of the effluent. Section 74 demands that effluent generated from any trade undertaking are discharged only into the existing sewerage system upon issuance of a licence from the Authority.

Environmental Offences recognized by EMCA
Offences as recognized by EMCA act 1999are outlined in the following sections
137 – Offences relating to inspection.
138 – Offences relating to environmental impact assessment.
139 – Offences relating to records.
140 – Offences relating to standards.
141 – Offences relating to hazardous wastes, materials, chemicals and radioactive substances.
142 – Offences relating to pollution.
143 – Offences relating to environmental restoration orders, easements, and conservation orders.
144 – General Penalty.
145 – Offences by bodies corporate, partnership, principals and employees.
146 – Forfeiture, cancellation and other orders.
147 – Regulations.
148 – Existing laws relative to the environment.

ENVIRONMENTAL   OFFENCES
137.
(a)   hinders or obstructs an environmental inspector in the exercise of his duties under this Act or regulations made thereunder;
(b)   fails to comply with a lawful order or requirement made by an environmental inspector in accordance with this Act or regulations made thereunder;
(c)   refuses an environmental inspector entry upon any land or into any premises, vessel or motor vehicle which he is empowered to enter under this Act or regulations made thereunder;
(d)   impersonates an environmental inspector;
(e)   refuses an environmental inspector access to records or documents kept pursuant to the provisions of this Act or regulations made thereunder;
(f)   fails to state or wrongly states his name or address to an environmental inspector in the cause of his duties under this Act or regulations made thereunder;
(g)   misleads or gives wrongful information to an environmental inspector under this Act or regulations made thereunder;
(h)   fails, neglects or refuses to carry out an improvement order issued under this Act by an environmental inspector;

commits an offence and shall, on conviction be liable to imprisonment for a term not exceeding twenty four months, or to a fine of not more than five hundred thousand shillings, or both.

138.
(a)   fails to submit a project report contrary to the requirements of section 58 of this Act;
(b)   fails to prepare an environmental impact assessment report in accordance with the requirements of this Act or regulations made thereunder;
(c)   fraudulently makes false statements in an environmental impact assessment report submitted under this Act or regulations made thereunder; commits an offence and is liable on conviction to imprisonment for a term not exceeding twenty four months or to a fine of not more than two million shillings or to both such imprisonment and fine.

139.
(a)   fails to keep records required to be kept under this Act;
(b)   fraudulently alters any records required to be kept under this Act;
(c)   fraudulently makes false statements in any records required to be kept under this Act;
commits an offence and is liable upon conviction to a fine of not more than five hundred thousand shillings or to imprisonment for a term of not more than eighteen months or to both such fine and imprisonment.

140.
(a)   contravenes any environmental standard prescribed under this Act;
(b)   contravenes any measure prescribed under this Act;
(c)   uses the environment or natural resources in a wasteful and destructive manner contrary to measures prescribed under this Act;
commits an offence and shall be liable upon conviction, to a fine of not more than five hundred thousand shillings or to imprisonment for a term of not more than twenty four months or to both such fine and imprisonment.

141.
(a) fails to manage any hazardous waste and materials in accordance with this Act;

(b) imports any hazardous waste contrary to this Act;

(c) knowingly mislabels any waste, pesticide, chemical, toxic substance or radioactive matter;

(d) fails to manage any chemical or radioactive substance in accordance with this Act;

(e) aids or abets illegal trafficking in hazardous waste, chemicals, toxic substances and pesticides or hazardous substances;

(f) disposes of any chemical contrary to this Act or hazardous waste within Kenya;

(g) withholds information or provides false information about the management of hazardous wastes, chemicals or radioactive substances;

commits an offence and shall, on conviction, be liable to a fine of not less than one million shillings, or to imprisonment for a term of not less than tow years, or to both.

142.(1)
(a) discharges any dangerous materials, substances, oil, oil mixtures into land, water, air, or aquatic environment contrary to the provisions of this Act;

(b) pollutes the environment contrary to the provisions of this Act;

(c) discharges any pollutant into the environment contrary to the provisions of this Act;

commits an offence and shall on conviction, be liable to a fine not exceeding five hundred thousand shillings.

142 (2)
In addition to any sentence that the Court may impose upon a polluter under subsection (1) of this Section, the Court may direct that person to –

(a) pay the full cost of cleaning up the polluted environment and of removing the pollution;

(b) clean up the polluted environment and remove the effects of pollution to the satisfaction of the Authority.

(3) Without prejudice to the provisions of subsections (1) (2) of this section, the court may direct the polluter to meet the cost of the pollution to any third parties through adequate compensation, restoration or restitution.

143.
(a) Fails, neglects or refuses to comply with an environmental restoration order made under this Act;

(b)   fails, neglects or refuses to comply with an environmental easement, issued under this Act;

(c)   fails, neglects or refuses to comply with an environmental conservation order made under this Act;

commits an offence and shall on conviction, be liable to imprisonment for a term not exceeding twelve months, or to a fine not exceeding five hundred thousand shillings, or to both.

144.

Any person who commits an offence against any provision of this Act or of regulations made thereunder for which no other penalty is specifically provided is liable, upon conviction, to imprisonment for a term of not more than eighteen months or to a fine of not more than three hundred and fifty thousand shillings or to both such fine and imprisonment.

145.(1)

When an offence against this Act, is committed by a body corporate, the body corporate and every director or office of the body corporate who had knowledge of the commission of the offence and who did not exercise due diligence, efficiency and economy to ensure compliance with this Act, shall be guilty of an offence.

(2) Where an offence is committed under this Act by a partnership, every partner or officer of the partnership who had knowledge or who should have had knowledge of the commission of the offence and who did not exercise due diligence, efficiency and economy to ensure compliance with this Act, commits an offence.

146.(1)

The Court before which a person is charged for an offence under this Act or any regulations made thereunder may, in addition to any other order:-

(a)   upon the conviction of the accused; or

(b)   if it is satisfied that an offence was committed notwithstanding that no person has been convicted of the offence;

order that the substance, motor vehicle, equipment and appliance or other thing by means whereof the offence concerned was committed or which was used in the commission of the offence be forfeited to the State and be disposed of as the court may direct.

(2) In making the order to forfeit under subsection (1) the Court may also order that the cost of disposing of the substance, motor vehicle, equipment, appliance or any other thing provided for in that subsection be borne by the person convicted thereunder.

(3) The Court may further order that any licence, permit or any authorisation given under this Act, and to which the offence relates, be cancelled.

(4) The Court may further issue an order requiring that a convicted person restores at his own cost, the environment to as near as it may be to its original state prior to the offence.

(5) The Court may in addition issue an environmental restoration order against the person convicted in accordance with the provisions of this Act.

(items in italics....SOURCE: EMCA 1999)

ROLE OF RECONNAISSANCE VS SURVEILLANCE IN ENVIRONMENTAL INSPECTIONS / ASSESSMENTS

Reconnaissance is an important and often overlooked step to pre-inspection preparation. Drive by and scout the facility before you attempt to enter. You may wish to consider the possibilities and adjust your approach. Take time to review your kit and checklists. The facility is never what you anticipated in the office. Use this time to adapt to any previously unforeseen contingencies. Consider the site layout, safety considerations, places and operations you want to include in your inspection, and decide if there is anything going on you want to visit immediately.

A reconnaissance over a longer period of time is called surveillance.

RISK ASSESSMENT AND HAZARDS IN ENVIRONMENTAL ASSESSMENTS

Essay: Risk Assessment (RA) of a facility with hospital and in a busy industrial area; 5 hazards to beware of:

FIELD SAFETY

Environmental compliance inspections are potentially dangerous. These dangers can be minimized through adequate knowledge and planning. There are many chemicals produced, stored, transported or used annually. Industrial sites producing or using these chemicals have process machinery, transporting equipment, structures and conditions that present their own hazards. The ultimate responsibility for your safety rests with you.

THE FIVE HAZARD CATEGORIES:
• Chemical • Fire and Explosion • Radiological • Biological • Physical

Chemical Hazards
Chemicals may be solids, liquids or gaseous. The health effects of chemical exposures may be either chronic or acute. Exposure may be direct or indirect. Reactions may be immediate or require long periods of time to manifest themselves such as with carcinogens. Health effects may be cumulative from many exposures over time. You must not depend upon your senses alone to warn you of exposure, as your reactions may not be quick enough to prevent injury or even death. Therefore you need to gather the necessary information, plan ahead and provide yourself with the correct personal protective equipment (PPE) and caution against all hazards before entering a potentially dangerous area.

Fire and Explosion
Fire or explosions may result from chemical reactions such as nitric acid and wood, sodium and water, aluminum powder and iron oxide. Even flour dust can explode under the right conditions. Combustion needs three things to take place. These three things are known as the fire triangle; and consist of fuel, heat, and an ignition source. The typical ambient breathing atmosphere has sufficient oxygen for combustion. In this case, all that is needed is an ignition source and fuel. There are many substances that can produce a fire or explosion that are found in industrial settings. Sometimes, generally safe chemicals can produce dangerous by-products. These may include peroxides, uncontrolled off-gassing, or combinations of incompatible materials that produce flammable or explosive mixtures. Changes in temperature may cause chemicals to boil causing a Boiling Liquid Expanding Vapor Explosion (BLEVE). Other ignition source may include camera, flashlights, and cellular phones.

Cigarettes are one of the most common accidental ignition sources in an industrial setting. Only smoke in areas designated for smoking or better yet, doesn't smoke at all. Health data indicates that all respiratory exposures to hazardous materials are many times more likely to cause damage if you smoke.

Radiological Hazards
Radiation sources may present external or internal danger. Some common sources are medical equipment, radioactive wastes from medical facilities, X-ray equipment, some electronic equipment, and even smoke detectors. Generally the greater the radiological hazard the more likely it is to be controlled. Highly radioactive sources will often have an obvious means of identification through hazard markers, labels or through detection equipment.

Biological Hazards
Biological hazards cause more lost man-hours for inspectors than all the other hazards combined. They consist of micro and macro-biological sources. Microbiological sources include viruses, bacteria and parasites. Every facility is a separate environment where plant personnel bring bacteria and disease into a central location. You should be particularly cautious around food and water sources, rest rooms and washing facilities. Macro-biological sources may cause harm from bites or stings and include things like guard dogs, insects, snakes and other animals. Biohazards also include botanical sources such as poisonous plants and allergic reactions caused by dust or pollen.

Physical Hazards
These include things that cut or crush you, things that you might trip over or fall into or slip on. They also include extremely high or low temperatures, dry or humid atmospheres, poor lighting and excessive noise. The potential of injury from physical hazards may be increased by circumstances where your senses are impaired, such as poor hearing because of hearing protection or an inability to communicate by voice because of excessive noise. Visibility may be impaired from a full-face respirator. Bulky protective clothing may make it difficult to move around in tight spaces. Protective clothing may be a hazard because it is too hot, heavy or bulky. There is a fine line between paranoia and

prudent caution but in the end it is always better to be cautious. A thorough and comprehensive understanding of real and potential hazards is best achieved by having a safety conscious attitude.

The following are some of the more insidious hazards that are often overlooked:

Oxygen Deficient Areas
These may exist in confined spaces and depressions. Oxygen can be displaced by other gasses or be consumed by chemical reaction. Excessive concentrations of oxygen can be dangerous because of increased risk of combustion or explosion.

Confined Spaces
These can contain pockets of trapped gasses. Alleyways between buildings are often over looked but may contain stagnant gasses and trapped fumes. Ditches and depressions may contain denser gasses such as methane, carbon monoxide or hydrogen sulfide. Trucks, railroad cars, and ship cargo holds may also trap dangerous gases.

Electrical Hazards
These may be obvious such as transformers, exposed wires or electrical panels. They may also include lightening or static discharges generated by high voltage electrical equipment. Underground electrical cables may be encountered during excavation. High voltages can "arc" if you provide a better electrical path to ground.

Fatigue and Stress
Fatigue and stress reduces sound judgment. Avoid excessive stress of all kinds. Stay warm and dry. Avoid extremes of heat or cold and provide adequate insulation. Monitor fatigue and allow adequate rest periods. Fatigue may also alter behavior and create tensions among the people you are working with. Sometimes wearing Personal Protective Equipment (PPE) causes physical or psychological stress. Provide shelter and a place to gather and organize your resources. Provide adequate, safe drinking water or liquid refreshment (not diuretics like coffee or tea). Keep high intensity work time to a minimum.

Loss of Peripheral Perception

This can result from focusing your concentration too closely is another common error. This may distract you from other dangers around you. In areas where hazards are high, use the "buddy system" to work in teams and watch out for each other.

KENYAN ENVIRONMENTAL LEGISLATION AND ITS IMPACT ON THE SOCIETY

This paper describes the policy and legal basis within which environmental legislation may be implemented. Regulations and standards applicable to the project are referred to. Environmental Impact Assessment is a tool for environmental conservation and has been identified as a key component in new project implementation. At the national level, Kenya has put into place necessary legislation that requires environmental impact assessment to be carried out on every new major project, activity or programme (EMCA, 1999), and that a report be submitted to the National Environmental Management Authority (NEMA) for approval and issuance of relevant certificates and/or licences.

To facilitate this process, regulations on environmental impact assessment and environmental audits have been established under the Kenya Gazette Supplement No. 56 of 13th June, 2003. Besides, a number of other national policies and legal statutes have been reviewed to enhance environmental sustainability in national development projects across all sectors. Some of the policy and legal provisions are presented in the following sub-sections.

COMPLIANCE INDICATORS:

Environmental Compliance and enforcement indicators is means of measuring the results of environment compliance and enforcement activities, both to reduce illegal activities, as well as to improve the ultimate state of the environment.

Indicators are a useful part of the compliance system, and offer a sound empirical foundation. Indicators are a method of displaying information about a complex phenomenon in a logical and concise manner that can readily be understood and communicated to other decision makers and other stakeholders. In the environmental context, indicators are used to measure the status of water and air quality, waste management and land use. They are an important part of a pragmatic, empirically grounded approach to environmental management based on the collection of hard data on actual consequences of decisions that then can inform subsequent rounds of decision making in a continuous information feedback loop that enables dynamic readjustment of policy and practice.

Currently, there are no comprehensive indicators of law and policy responses to environmental problems, especially those relating to compliance and enforcement. In response to this demand, International network for environmental compliance and enforcement (INECE) launched a project at the 2002 world summit on SD to create a framework for developing indicators to measure the effects of compliance and enforcement activities in the quality of the environment. Output indicators measure government activities, work products or actions e.g. number of enforcement cases settled per year. Intermediate outcome e.g. the amount / pounds of pollutant/ air pollutant reduced per year as a result of compliance and enforcement activities- to measure their progress towards achieving a change in behaviour or knowledge. It is anticipated that these will be used to determine final outcome indicators that measure the results of compliance promotion and enforcement actions on the state of the environment (i.e. improvement of water quality, air quality etc.

The indicators are also used to determine how environmental compliance brings countries closer to achieving the MDGs and other development objectives. In a region where effluent quality standards are to be introduced for the first time, discuss the relative advantages and disadvantages of setting:
(i) National emission standards (The uniform effluent approach); and
(ii) Ambient standards for individual works (the water quality objectives approach)

The broad objective of regulatory instruments in pollution control is protection of receiving environments and public health. This is to ensure that effluents do not cause acute or chronic damage to the biotic and abiotic forces. Regulatory instruments involve deciding and enforcing measurable and achievable effluent standards. If well constituted, standards can be used to determine effluent fees, therefore acting as a way of sustaining, running (operation and maintenance of) and expanding the facilities. However, in systems where they are new, a lot of care needs to be taken, as the whole process may cause unnecessary tension if the set standards, using whichever method, are neither enforceable nor achievable.

The water / effluent quality assessment instrument should have the capacity to measure the basic standards parameters, and interpret them accordingly with the specific setting in mind. The use of standards is generally referred to as the command and control strategy. It is generally known, from experience, to be cumbersome where there are no facilities and institutions to support, develop and enforce. An incremental command and control approach would be a compromise strategy for poor developing countries whose immediate priority is poverty alleviation, employment, and food production (all of which demand investment, sometimes even polluting facilities). Setting very stringent standards may in such cases deter development (industrialization and commercial production processes); such countries may not consider environmental conservation a priority, since the level of pollution is still negligible compared to the gains associated with the pollution facilities.

Approaches for meeting effluent quality standards are: (i) Ambient standards (Water quality objective) (ii) Emission standards (Uniform effluent approach). The following table gives a comparison between these two approaches:

	Advantages	Disadvantages
Ambient standards approach	Focuses on the Water quality as an objective Suitable to the local environment; Takes into account local factors; Flexible and therefore more appropriate; Less capital cost; Varies with flow and final water use; Case-specific and therefore more appropriate in protecting the ecosystem; More participatory as decisions made more often; Timely self cleansing by receiving waters possible; Takes into account use of the receiving water body;	Difficult to measure correctly; Amenable to abuse; Needs regular field assessment to justify change of standards; Amenable to controversies; Can be improved on incremental basis; Does not consider water use;
Emissions standards approach	Focuses on uniform effluent approach Effluent of same standard; Less discriminative; Considers actual and potential water use; Uniform quality monitoring facilities; Easy to monitor and evaluate;	Difficult to enforce uniformity; Less participatory/ more alienating; Heavier capital outlay; Standards may be inappropriate for some situations e.g. during low flow seasons and times of the day e.g. nights); Needs a lot of baseline data; Likely to cause serious ecological problems; Difficult to agree on common standards to use;

Water quality standards (Afullo, A 1995). (Source: numbers in brackets)

Parameter	Irrigation	Drinking	Surface water discharge
Dissolved O2	3(5)	6 (5)	-
Boron (mg/L)	3(2,3,6)	1(4,7)	-
PO43- (mg/L)	-	6(5)	-
NO3- (mg/L)	30(6,9)	20(4)	50(5,7)
NO2- (mg/L)	-	0.2(4)	1(5)
S2- (mg/L)	0.1 (5)	0(4)	0.05(5,7)
EC(uS/cm)	.2-.25(1,3,9)	0.4 (4)	-
TDS (mg/L)	.2-.5 (1,3,9)	0.5 (70)	-
TSS (mg/L)	-	0 (15,7)	30 (6)
BOD5,25oC	11 (5)	6 (5)	20(6)
Na (mg/L)	50 (3)	200 (4)	-
K (mg/L)	-	12 (4)	-
Ca (mg/L)	-	75 (7)	-
Mg (mg/L)	-	50(4)	-
HCO3- (mg/L)	91.5(6)	-	-

1. Wilcox and Durum (1967). 2.FAO/UNESCO (1973). 3. Ayers and Westcot (1976). 4. WHO (1984). 5. WHO (1963). 6. UCCC (1974). 7. Kenya Bureau of standards in: (Nyamu, 1986). 8. GOK, (1983/4). 9. Ayers and Westcot (1985).

Introduction

Environmental Indicators are environmental parameters that represent the conditions of a given situation. Indices are qualitative (descriptive) or quantitative (numerical) categorization of large numbers of environmental parameters (Canter (1996); Connel and Miller (1984)). They can be compounded into a single value. The use of both Physico-chemical and biological parameters complementarily gives more monitoring scope than using one. (Connell and Miller (1984). Generally, monitoring can take two forms: (i) direct measurement of concentration of pollutants or of key substances depleted by pollution (e.g. O_2). This may be useful in monitoring the aquatic environment and fisheries. (ii) The use of biological indices which range from bioassays with micro-organism and BOD measurements to the kind of total community indicators e.g. plants to monitor the soils and water conditions (Kapoor, 1989).

(i) Water Quality monitoring in Environmental Health

In environmental health, the key concerns are biological, microbiological and chemical composition of water (in this case, the river water, waste-water and groundwater). The bacteriological quality of water is normally the key water quality system. The index used here is called the Coliform index. It is a measure of the concentration of Coliform bacteria or E. coli in a water sample. Coliforms are bacterial mostly associated with human feaces, but can also be found in feaces of other animals (eg cattle). E coli is bacteria only found in human feaces, and therefore confirms faecal contamination of water. It is measured using the Most probable number (MPN) which is the bacterial density which if it had been present in the sample under examination, would more frequently than any other, have given the observed analytical results (Kapoor, 1989). E coli is a bacterium that, if present in a water sample, would confirm the contamination of water source by human feaces. This would straight away indicate possible risks of cholera, typhoid, paratyphoid, amoebic dysentery, infective hepatitis /jaundice gastroenteritis / diarrheas and intestinal worms to users of the water source.

However, as Big town grows, industrial effluents will increase, and some might be disposed of into the water. Industrial effluents have heavy metal (Lead, Manganese, Arsenic, Chromium), pesticide, and other anions such as nitrate, cyanides and fluoride that directly affect human health. Heavy metals and pesticides may be carcinogens, mutagens, or teratogens or cause brain disorders. Fluorine causes Fluorosis and dental caries etc, while nitrate causes methaemoglobinaemia, and some schools of thought have associated it with early onset of hypertension (Afullo, 1995). In respect of the many parameters useful in environmental health, I would propose a more comprehensive weighted index, giving the following weights to each parameter: E.coli count (40%), other pathogens (20%); heavy metals (15%); pesticide (10%); others e.g. TSS[1] (15%)

(ii) Fisheries Water Quality
For Fisheries, the most important parameter, which can also act as an indicator is the Dissolved Oxygen (DO). Action / intervention level is 3ppm[2], but threshold value is 5ppm (Kapoor, 1989) (Higher value is preferred). If DO level falls below 4 ppm, the fish and other aquatic life will start dying (Kapoor, 1989). In fresh water at 200, 250 and 300, the Dissolved Oxygen concentrations are 9.2%, 8.4% and 7.6% respectively. The levels fall with increasing salinity of water (i.e. as the TDS[5] increases).On the other hand, the carbon-dioxide concentration should not be more than 40 ppm (action level), while the threshold level is 20 ppm (Kapok, 1989). The DO is an indication

of BOD[3] and COD[4], thus making it more inclusive. The DO can be used to calculate the Oxygen balance (OB) of a water body. OB is the relationship between the BOD of sewage effluent and the oxygen available in the diluting water. It is a measure of the amount of sewage that a given watercourse has capacity to handle.

(iii) Agricultural water Quality

Agricultural water quality is determined mainly on basis of effects on the crop, and effects on the soil. Reclaimed effluent can be used in agriculture as long as the BOD and TSS are reduced by about 85%, the effluent chlorinated to a residual before application to the farm (Kapoor, 1989. Another view is that a well-oxidized, non-putrescible and disinfected or filtered effluent should satisfy the following conditions: In 20 consecutive samples, from each of which five 10 ml samples are examined, not more than 10 samples shall be positive for the members of coliforms group (Kapoor, 1989). However, all this depends on which crops are grown. Salad crops and those likely to be eaten raw should not be produced with bacterially contaminated water. There are many other parameters that should be considered alongside the bacteriological water quality. These relate to parameters such as salinity, cations, anions, divalent cations, Sodium, Boron, Selenium, and Nitrates.

Salinity
TDS and / or EC[6] of water is a reflection of salinity of water, which has serious effects on the crop.. Saline waters cause osmotic pressures (OP) & imbalances and water stress on plants, thereby making them to wither, wilt or die. Thus 1000 ppm is the most appropriate irrigation water salinity (can be considered target level); 2000ppm is suitable for most of the plants except salt-sensitive ones (can be considered threshold level); while 3000 ppm is the action / intervention level. **According to Ayers and Westcot (1976), OP   =   -0.36 X   EC**
Where: OP= osmotic potential or pressure in bars; EC = electrolytic conductivity in mmhos/cm; -0.36 is the conversion factor.

Irrigation water Salinity indicators
Many indicator crops are used to monitor EC changes in irrigation water. These include: Beans (2-12 mmhos/cm); Corn (4-20 mmhos/cm); Alfalfa (6-30 mmhos/cm); Date palm (10-63 mmhos/cm) and Grain barley (20-56 mmhos/cm) (Ayers and Westcot, 1976)

Ionic balance of irrigation water (ESP and SAR)

Cation and anion balance and ratios give the best indication of irrigation water quality, as they have serious effects on the long-term soil structure and fertility. Indices such as Exchangeable Sodium percentage (ESP) and Sodium Adsorption Ratio (SAR) have been commonly used to assess irrigation water suitability. An ESP figure less than 60%, and an SAR figure less than 10% are most preferred (Kapoor, 1989; Ayers and Westcot, 1976).

$$ESP = \frac{Na^+ \times 100}{Na^+ + K^+ + Ca^{2+} + Mg^{2+}}$$

{NB: The figures represent concentrations of individual elements; each expressed in milli-equivalents / litre (me/l)}

$$SAR = \frac{Na}{\left[\dfrac{Ca^{2+} + Mg^{2+}}{2} \right]^{\frac{1}{2}}}$$

Where: Na = Sodium; K= Potassium; Ca = Calcium; and Mg is magnesium.

Higher values of ESP and SAR than the above are detrimental to soils. Sodium is the basis of comparison with either all major cations (in the case of ESP) or divalent cations (SAR) because it deflocculates soils. It causes separation and dispersal of particles, making the soil loose and amenable to erosion.

Boron: Boron is another key parameter used in assessing irrigation water quality, as high levels are injurious to plants. 3 ppm is the action level, I mg/l is threshold level, while anything less is good for all crops.

Nitrate (NO_3^-): NO_3^- level in irrigation water is vital because (i) reasonably low levels of 40 ppm and less facilitate plant growth, as nitrate is an essential plant macro-nutrient (ii) NO_3^- levels above 40 ppm cause eutrophication of water bodies, and is undesirable (Afullo, 2003, 1995).

CUMULATIVE ENVIRONMENTAL EFFECTS
Introduction:
Cumulative effects are those that are not necessarily directly attributable to a specific project alone. They may be additive or synergistic (i.e. more than simple interaction with other actions / projects); or may affect an area wider than the project coverage area (spatial), and may spread long after the project

implementation (temporal). Some possible cumulative effects of this project are:

1. Increased concentration of micro biological pollutants (e.g. E coli loads) as the sewage from the town are released into a river with reduced flow / discharge (i.e., higher concentration of pathologically active pollutants in the river).
2. Increased concentration of inorganic industrial pollutants (e.g. sulphates) industrial waste from the town are released into a river with reduced flow / discharge (i.e., higher concentration of inorganic pollutants in the river).
3. Increased Biochemical Oxygen Demand (BOD) and Chemical Oxygen Demand (COD) as raw organic-matter-rich sewage and industrial waste from the town are released into a river with reduced flow / discharge (i.e., higher oxygen demand in the river).
4. The new humid microclimate is likely to increase fungal diseases, which will necessitate increased use of fungicides. This will increase river pollution, where the fungicide residues may have additive / synergistic effects. This may potentiate river pollution many-fold.
5. The project is likely to affect agricultural, land use, land tenure and other sector policies and patterns if it properly integrates CPP, and takes a strategic approach.
6. The area is likely to acquire a new urban status after the project completion, leading to a new cash economy and related urban characteristics.

Benefits of using strategic environmental assessment (SEA) techniques

Strategic Environmental Assessment (SEA) is an improvement of the formal Environmental Impact Assessment (EIA) involving the analysis of policies, plans and programs. The EIA only deals with project-based impacts. Thus EIA is short-lived, and addresses only issues directly relevant to the project, within its cycle from conception, up to evaluation. Secondly, the EIA mostly stops immediately after project approval by the relevant authority i.e., at decision-making stage. SEA addresses cumulative (additive and synergistic) effects of plans, policies and programs in a wider framework that would also cover project-based assessments. SEA sets the legal, institutional, economic, social and environmental framework for broader operation of EIAs ((Lee and George (2000), Dijkstra (2003) and Canter (1977)).

Due to these limitations of the EIA, SEA strives to fill some of the gaps. The SEA approach deals with impacts in a wider scale, and covers more time; it is a continuous process as the assessments are part of an institution's policies, plans and programs. Generally, projects, programs, plans and policies have varied scope, with the project being the lowest and policy is the highest. The relationship can be illustrated as follows:

A policy may cover a whole country or region, and therefore tends to have a cumulative (i.e., temporal and spatial) consideration of impacts. Project-based EIAs, on the other hand, address narrower site-based or local impacts (canter, 1996). Since the policies are wider in scope, they tend to cut across many projects, plans and programs. On the other hand, a program covers a number of projects, while a plan covers a number of programs. Therefore program, plan and policy-based environmental assessments are relevant to and have the capacity to cover more projects than the project-based assessments (Dijkstra T (2003).

When SEA is embedded in a country's legislature, it has the capacity to reduce costs associated with project assessment and costs, since some impacts assessed from earlier projects within that wider framework can be applied to a number of projects; this caters for a wider networking. Similarly, through SEA, broad guidelines can be institutionalized to ascertain minimum standards of the environmental assessment. SEA, being wider in scope, is likely to cover a wider area, and can therefore have more capacity to handle impact assessments, which gives it more credibility (Dijkstra, 2003). SEA also facilitates decision-making (especially once the analytical environmental assessment is incorporated), thereby reducing project delays and harmonizes the entire project cycle, so that the direction of any project is certain depending on its nature, location and scope.

EA can give earlier guidelines, going by the experience from related or similar projects, on what EIA steps can be skipped, and which are mandatory. This saves time and resources, eases the screening and scoping stages, and makes it possible for the other stages of the EIA to be applied in time from the project conception through to the end. This is likely to provide some social and economic gains to the community, and improve the general standard of living since the reduced cost of project implementation may benefit the users of the product or service from that project.

Lastly, due to clear inscription of environmental assessment guidelines in national and regional records, wider participation in the process is likely, as the basis of each project will already be clear to a section of the stakeholders. This can act as a conflict-resolution strategy. However, these assessment guidelines can be availed in local languages by print and alternative media to facilitate wider participation, since in the developing countries, the EIA is either new, or non-existent (Lee and George, 2000).

The SEA can therefore be a reasonable substitute for EIAs, a role they can perform even more effectively. In Botswana, for instance, EIA is not yet mandatory (GoB, 2003). Yet the number of new projects being developed is so many that unless their long-term repercussions are addressed early through alternative means such as SEA, there projects can only be time bombs.

A typical methodology for health impact assessment is the health risk assessment and management.
Health risk assessment is a strategy meant to identify health hazards, the primary receptors, routes of exposure, the nature of damage (i.e. risks involved), and instituting control and mitigation measures (CIEH (1998), Afullo (2003), Afullo (2004)). It uses the same guideline as the environmental or social impact assessment, but focuses more on the human environment and health implications.

In general, risk assessment and an environmental health assessment method observes three main stages:
(i) Process / activity or action causes a change in the physical environment;
(ii) Change may cause an impact, depending on the threshold and vulnerability of the receptor and
 (iii)The impact may necessitate a mitigation process.

Accordingly, if the health impacts are too significant, an alternative may be sought, which may again take the process back to stage 1 (activity / action / process. Changes will always occur, but only warrant mitigation in highly vulnerable members of the society e.g. expectant mothers, children, orphans, women, street children, etc. The guidelines are adapted from CIEH (1998) and Dijkstra (2003)).

PREVENTION OF ENVIRONMENTAL DEGRADATION: EXECUTIVE SUMMARY

One universal global concern at the moment is environmental degradation (ED). It results from extraction of environmental assets, processing or fabrication activities or product consumption. The extractive processes lead to resource depletion, while processing and consumption both cause pollution. Pollution is the decline in quality of a resource following disposal of more substances (wasted matter and / or energy) than its resilience or assimilative capacity. Resource depletion, on the other hand, is the decline in the quantity of a resource. Therefore, a resource may be degraded by decline in quality, quantity or diversity.

ED has serious repercussions. It negatively impacts man in terms of health deterioration and increased medical bills, lost productivity, cost of environmental restoration, increased input costs following decline in supply, increased cost of input purification, elevated global temperatures etc. Mankind incurs extra costs arising from environmental degradation over and above the private costs of products- called externalities, which arise mainly from use of common access resources. Environmental assets face dangers of degradation mainly due to the individuals' insatiable urge to maximize utility, but in a competitive way. This leads to the tragedy of the commons. Therefore, some level of control of resource use is vital in protection of environmental assets.

Resource use control is meant to assign property rights to users. It may take the form of standards (also called Command and control), economic instruments or moral persuasion. Standards are easy to find from literature, but difficult to implement, besides it not offering incentives to polluters to reduce pollution. Economic instruments include tradable pollution permits and pollution charges. The latter is determined based on the equilibrium between marginal abatement cost and marginal damage cost. It gives polluters a number of options, but most importantly, encourages them to invest in pollution abatement technologies. Moral persuasion involves public education and awareness, with all stakeholders being given information with the aim of empowering them to be custodians of the environment.

Generally, a combination of methods is used. Whatever the mix, however, moral persuasion should be included, for it is a means of enlisting support of everybody, including the polluters in a positive way in the broad perspective of primary environmental care. This can ensure sustainability of preventing ED.

INTRODUCTION

The environment is the total sum of all natural resources-including biotic (living) and abiotic (non-living) factors (Afullo, 2003). They interact through complex networks so that one factor, regardless of size, plays a role in keeping the total equilibrium in place. The biotic factors referred to here include man, all living plants, animals, micro-organisms, etc, while the abiotic factors include rocks, the planets, air and all its constituents, energy, water resources, soils, etc. The members are in continuous interaction, and together they form an independent, self-sustaining system called the ecosystem. The continuous interaction within the ecosystem assures some level of stability. Matter and energy are continuously lost and gained, giving a general balance or equilibrium. They are the natural (environmental) assets which man currently calls natural resources.

Through exploitation of the natural resources, man has gone to some extreme, leading to the destabilization of the ecosystem. This has come through excessive extractive processes (of minerals, water abstraction) - hereafter called depletion. The resources are reduced in Quantity beyond their regenerative capacity. Alternatively, man may cause degradation by releasing too much matter (industrial effluents) and / or energy (hot gaseous outputs or effluents) into parts of the ecosystem-air, water, land, or soils. This latter destructive waste of environmental assets is called pollution.

ENVIRONMENTAL DEGRADATION

Environmental degradation (ED) is therefore a concept comprising two major components - depletion (an extractive process) and / or pollution (a release / resource contamination / wasting process). Most of the time, the extractive aspect of environmental degradation utilize environmental assets in a wasteful way, leading to release of a lot of "Waste" which is then disposed of in parts of the ecosystem which may not be ready to accommodate them due to wrong disposal timing, unbearable quantities released, or their incompatible nature with the receiving body hereafter called the sink). It may involve one or a combination of reduced quantity or quality of an environmental asset. ED can take many forms. These include decline in quantity, quality or diversity. Therefore efforts to prevent ED revolve around maintaining these aspects. An illustration is given below to explain the concept further (Dijkstra, 2003).

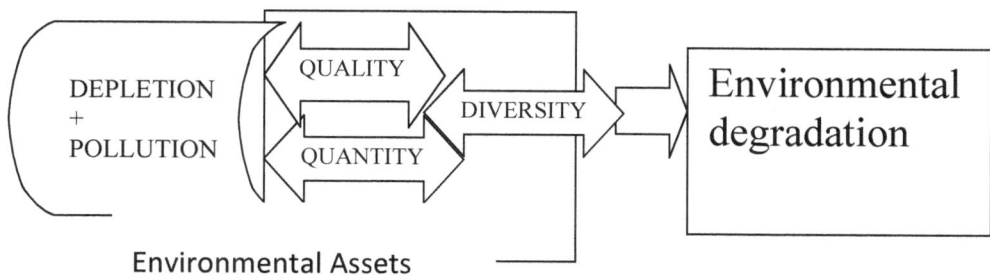

Figure W: Degradation of environmental assets.

Tragedy of the commons

Commonly shared resources may be either open access resources (OAR) or common property resources (CPR). OAR and CPR are essentially the same except in the latter case, there is controlled use so that assess is not open, and there is some exclusivity (The users may buy rights to use the resources (e.g. in cases of cooperative societies where members make payments). In cases of an Open access resource, the utility derived by the next user is lower compared with that derived by the earlier users, i.e. there is some element of rivalry or competition.

The quantity or quality of the open access resource diminishes with use. This is worsened by the fact that there is no control of users, i.e. there is no exclusivity of users. It is in such situations that tragedy of the commons comes in, leading to complete depletion of the resource, especially non-renewable ones. This unsustainable resource use may also affect renewable resources if the rate of resource recovery (renewal) is slower than the rate of extraction. They may slowly be depleted. It is therefore vital to introduce some control in use of environmental assets.

In a typical situation of tragedy of the commons, individual users aim at maximizing their utility of the common resource, and this introduces some level of competition. The resource diminishes, mostly to the detriment of all. The cost of resource damage is shared by all the users, including the one who might have done the last activity that took the damage level to its threshold. Yet this last user (perhaps the one who benefited most) created a cost which (s)he only bears a fraction of, and is a cost which is lower than the marginal benefit (return) he got from the last action. Eventually, the common resource gets finished (depleted) and all suffer, or get alternative coping strategies. However, the satisfaction members derive from the use of the common resource can be increased if some rules are introduced. These are given below.

RULES GOVERNING SUSTAINABLE UTILIZATION OF ENVIRONMENTAL ASSETS

On the other hand, there is need to control the use of the environmental assets which eventually are the inputs from which pollutants come. Without property rights allocation, resources are bound to continue being degraded. The type of control mentioned here therefore is to do with assigning of property rights so that there is exclusivity, and with this will come responsibility and accountability. Introducing some level of control in form of property rights in use of the resources can control the extraction processes, and therefore save it from depletion. This may also save the users, as they will be guaranteed of availability of the resource long into the future, while at the same time they continue deriving optimal utility from using it.

The aim of sustainable natural resource management is to ensure that they their current use does not compromise the ability of future generations to use them, i.e. there should be a satisfactory level of utility to be derived from the use of the resources by the future generations. To ensure this happens, there needs to be some conditions governing the utilization of the resources. These include:
(i) Define clearly the boundaries of the resource;
(ii) Define clearly the users, i.e. there should be no open access use. This exclusivity brings in some kind of value (price) to the resource, unlike in a situation where the resource would be free access, and tragedy of the commons would ensue;

(iii) There should be rules governing the use of this environmental asset;

(iv) The rules should be made and decided together by all users in an open way to give them credibility and acceptability;

(v) There should be ways of monitoring the use of the resource, so that there is some accountability by every member;

(vi) There should be ways of punishing the users who break the rules of using the resources;

(vii) There should be some means of arbitration in situations of an impasse; this may create common grounds for consensus building;

When these guidelines are followed, there is bound to be transparency and responsibility in the utilization of environmental assets.

RESOURCE EFFICIENT TECHNOLOGIES (INPUT-OUTPUT BALANCE)

Once resource extraction is controlled, depletion may be reduced, and therefore environmental degradation will be equally less. This comes as a result of reduced inputs, which are acquired at a cost (membership fee or lease payment in case of land). Due to the cost of acquiring the environmental asset (input), the users may be more responsible and efficient in using it, thereby saving in overall production costs. Process efficiency can be introduced by investing in efficient technologies that produce less waste. This can only happen if there is a limit to the use of the input.

Mechanisms to improve production efficiency include concepts such as Best Available Technology Not Entailing Excessive Cost (BATNEEC), etc. However, using external forces best reinforces the incentive to invest in more efficient and resource saving technologies. Any technology that does not use inputs efficiently (i.e. wastes inputs) obviously produces more undesirable process products (wastes or effluents).

These wastes have to be released somewhere- in a part of the environment. The most common sinks are land, soil, or water. Land has acted as a sink for mostly solid wastes; Oceans have been the unwilling recipients of both solid and liquid wastes for many centuries; lakes have been sinks to millions of liters of liquid wastes (effluents); the atmosphere / air has been a silent receiver of waste gases since time immemorial. These sinks have for long been viewed as public resources. Their capacities have long been exhausted, yet may are still receiving wastes. It helps the resource users (and therefore waste generators) to save on direct production costs. However, this introduces other costs commonly borne by the public. These are the externalities, sometimes called social costs.

EXTERNALTIES
These are costs borne by the public, and mainly arising from environmental pollution. Externalities can be Reciprocal or unidirectional. Reciprocal externalities are those where all parties having rights of access to a resource are able to impose costs of damage on each other. This is applicable in situations where there is exclusivity, and some element of participatory management of the resource holds. In Unidirectional externalities, the generation of pollutants or emissions by one agent causes costs or consequences to others without being reprimanded. This is where the government can come is, as it applies to open access resources where there is loose control, if any. Both reciprocal and unidirectional costs may give rise to cost shifting or displacement costs, and may require different strategies. These approaches are discussed later under policy instruments to ED control.

NIMBY AND NIMTO SYNDROMES
There are various ways of managing undesirable production products (wastes). For the majority, the waste is a menace only if it is located in their property, vicinity, etc. It is therefore common for solid wastes to be removed from a yard, and thrown outside the yard (e.g. in a street). For this person, things are okay as long as the waste is out of their property and sight. This is what is called Not In My Backyard (NIMBY) Syndrome (Afullo, 2003).   On the other hand, the wastes in such public utilities such as streets are rarely, if ever, attended to by volunteers (unless there is a very strong ethical and / or financial motive). We view the environment as an acceptor of wastes with unlimited capacity. Yet our individual instincts make us see the mess caused by wastes. We, however, hope the next person sees the mess. And they do.

There is no ethical motive to control the wastes. Because I am not employed to do the job, i.e., it is Not In My Terms of Office (NIMTO) (Afullo, 2003). In such cases, we see financial motive as the overriding factor in waste management, even if our conscience is disturbed by the poor waste disposal. We look forward to somebody employing us, or compensating us for the contribution in clean –up campaigns. This is why most litter picking campaigns are highly publicized so that some well wisher can come in and make financial compensation (i.e., provide sponsorship to the activity), because waste management is not/ our in my terms of office, yet I / we volunteered and did it. This aspect of attitude will be mentioned again later under persuasion approach to environmental degradation control. Yet even as our conscience betray us with wrong financial motives to environmental pollution control, we bear the cost of pollution in form of externalities.

COST OF POLLUTION
Wanton waste disposal poses serious environmental health, ethics and aesthetic concerns. The environment absorbs the cost of pollution indirectly, as illustrated in the following ways:

(i)   Reduced Fish stock, leading to reduced supply / scarcity, leading to increased market prices. The willing fish consumer pays this higher price. The difference between the increased price and the original fish price (ceteris paribus) is the cost of aquatic resource pollution borne by the fish consumer.

(ii)   Gaseous emission to the atmosphere introduces toxic substances which cause respiratory diseases such as bronchitis, asthma etc. The human victims have to pay for treatment of these diseases, which came not because of a reason of their own making; they are paying the pollution costs.

(iii)   Gaseous emissions to the atmosphere from a pulp and paper industry contain acidic gases which, when moistened by the atmospheric water vapor, produce acid fumes which accelerate rusting of roofs, and wear and tear of some metallic structure. This increases depreciation costs of man-made assets (depreciation costs), as well as increase cost of repair (operation and maintenance). This is cost of pollution transferred to the innocent member of the public.

(iv)   Industrial effluent emissions to a lake increase concentration of undesirable components, leading to the need for treatment of the water before use. If it was already being treated, the effluents increase the need for more rigorous treatment – up to and including investing in some new more expensive water treatment technologies.

(v)   Disposal of domestic wastes to a surface water increases the pathogen load, leading to the unsuspecting low-income community downstream getting water-borne diseases, for which they have to seek treatment, or die from, or miss work for some days. Either way, they are bearing the burden of pollution.

Pollution has therefore long gone beyond threshold level in most environments. The best policy is to make the polluter pay (i.e., PPP) to avoid transferring the pollution cost to the rest (Kgathi and Bolaane, 2001). The resources such as oceans, lakes, land soils etc can no longer be treated as public or open access resources. There must be control in the way they are used. It is in light of this that a number of instruments have been introduced by different economies to contain the activities of polluters. Various mechanisms have been used to try to control ED. These may be economic, persuasion, or command and control (use of pollution standards). The economic instruments of pollution control include taxes and tradable pollution permits.

MARKET FAILURES

Market failures arise if the market price of a commodity only reflects private costs of production, but not social costs. This may be controlled by introducing charges such as tradable permits, regulations and polluter taxes. Thus the polluter pay policy (PPP) is introduced and enforced. Secondly, market failures may arise from the fact that some products have no markets at all, and interventions are required to establish prices for them. This may be approached by assigning property rights to the "commons", i.e. common access / open access resources. When market failures arise, government interventions can be useful. The intervention should ideally lead to a method of allocation that is more favorable than free market system. This was discussed earlier under tragedy of the commons.

POLLUTION CONTROL POLICY INSTRUMENTS

Pollution control policy instruments differ primarily on whether they are price-based or quantity based. Price-based system sets a direct price on behavior (legal or illegal behavior). These include taxes on emissions, fees for hazardous waste disposal or fines for non-compliance (Kgathi and Bolaane, 2001). Quantity-base instruments set direct restriction on volumes of inputs, emissions, or technologies. They may be both enforced and price-base instruments such as fines. Price instruments are very influential instruments. If resource prices are set too low e.g. user fees for forest products, excessive use will be made of them to depletion point. To ensure efficient use of resources, products should be priced at their marginal social cost. This comprises marginal cost of production plus external cost of pollution or resource degradation caused by the production of the good. If market functions well, prices may be able to reflect the marginal social and private costs of production. However, there may be market failures. These have been discussed above, and the policies described below are meant to cater for such failures. Lee and George categorize the instruments into three categories. These are command and control (C&C) instruments, Economic instruments and planning instruments (Lee and George, 2000).

REGULATION STANDARDS OR COMMAND & CONTROL (C&C) POLICY
This involves the introduction of a mechanism of monitoring the concentrations of some parameters of effluents from different sources, and setting a threshold level of pollution. Threshold level is the highest acceptable concentration of pollutant that which if exceeded by the generator, warrants payment of pollution fee. Thus there is a fixed charge for all pollutants regardless of the magnitude of pollution they cause. This may work well for areas where clean technologies are used. Different polluters produce different quantities and concentrations of wastes, making it very unfair for those contributing only marginally to pollution. Due to economies of scale different industries and polluters enjoy, the unit cost of pollution control (marginal abatement cost) vary with different polluters. Since the pollution charge may be small (and it normally is) the polluters would rather pollute, save the cost of pollution control, and get higher profits. These end up extracting more resources, and use them inefficiently (waste), and these undesirable products (wastes) are disposed of haphazardly. This situation does not offer any incentive for polluters to control pollution. Secondly, it is unfair to some polluters who have to pay the same pollution charge as the large-scale polluters. In the long term, it can encourage more to pollute.

According to lee and George (2000), C&C instruments take the form of permits and authorization procedures relating to:
(i)   The types of products that may be produced and used;
(ii)   The types And quantities of raw materials that may be abstracted and used for production and consumption;
(iii)   The technologies by which goods and materials may be produced;

(iv)   The maximum quantities and types of residuals which may be released into the environment;

(v)   The locations at which resource abstraction, production and other activities may take place.

MARGINAL ABATEMENT COST (MAC)

MAC is the cost of controlling an extra unit of pollutant / emission. This varies from polluter to polluter. In the illustration below, it would be unfair for firm B with low MAC (MAC_B) to be charged on the same scale as firm A with higher MAC (MAC_A), as they would not be encouraged to control pollution. With the low MAC for firm B it should be encouraged to abate pollution, thus helping reduce environmental degradation. But if there is no such an incentive, which command and control system does not offer. Since it has to pay the tax, whose level is much higher than its MAC, it would be tempted to perhaps pollute more (become more inefficient) as the charges for pollution are fixed, regardless of scale.   On the other hand, Firm A will find it cheaper to pollute since its MAC is higher than the tax. So it will prefer to pay the tax. There will be no incentive to invest in pollution abatement technologies. This is why this system of command and control does not work.

The second weakness of Command and control policy is that most of the time, standards for emissions and effluents are set based on those of the developed countries (Dijkstra, 2003; Kgathi and Bolaane, 2001). Their use requires serious consideration; at times decision making based on them may have to be discretionary. In other words, they need to be used with a lot of care, as the unique environments in the developing countries may still accommodate less stringent standards. At times, the standards adopted by the low-income countries are copied direct from a model high-income country that may have stopped using those standards themselves due to serious discrepancies. The time lag between creation of standards in a developed country and their implementation / adoption in the low-income countries may make the standards difficult, if not impossible to enforce. They may have been overtaken by events.

Thirdly, in command and control, effectiveness is low because monitoring and enforcement may be completely inadequate. Even in situations where monitoring may be feasible, it may be impeded by other unique environment-specific parameters related to culture, administration, manpower, and related budgetary constraints. At times, even corruption comes in to derail the whole exercise.

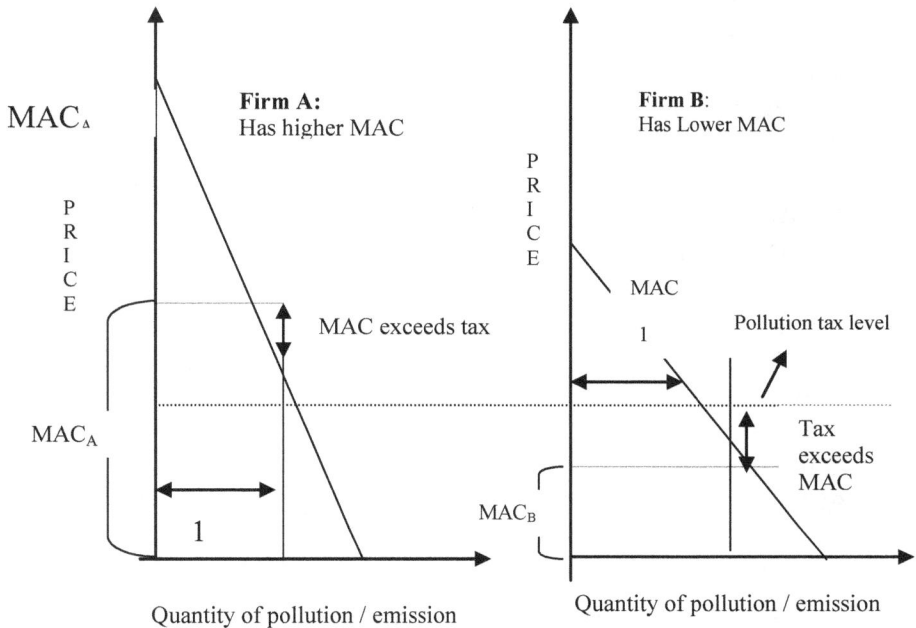

Quantity of pollution / emission

Quantity of pollution / emission

ECONOMIC INSTRUMENTS OF POLLUTION CONTROL

Due to the weaknesses of the command approach, there is more inclination towards market-based approaches (Kgathi and Bolaane, 2001). These are more flexible and efficient. At times, the trend is to implement market-based instruments to complement the command and control method. Economic instruments provide flexible options for selecting optimal pollution levels. They provide continuing motivation to reduce pollution below the levels set by regulators. They also encourage new pollution control technologies, production processes, and new non-polluting product. This provides incentives for research and Development (R&D).

Economic instruments influence behavior to improve performance through the use of :
(i)   Pollution charges and environmental taxes;
(ii)   Environmental protection subsidies and grants;
(iii)   Market creation schemes, such as emission trading schemes;

(iv)   Environmental licensing charges and fines for non-compliance with environmental regulations (Lee and George (2000))

Lee and George (2000) lists the following in the planning and other instruments category: Environmental planning studies; Environmental assessments strategic environmental assessments, Environmental impact assessments, and related measures; Environmental audit procedures and environmental management systems; and voluntary agreements to encourage compliance with environmental quality targets through such measures as industry covenant.

EFFLUENT OR EMISSION CHARGES

These are sometimes called user or product charges. They are levied on the amount of pollution that a polluter discharges, and are a direct method of putting price on the use of the environment.   They are very useful in discouraging pollution activities, processes, or inputs Dijkstra, 2003). They may also involve providing finances for pollution reduction. They are useful in controlling point source pollution, i.e. where emissions are from one identifiable source, and its quantity and toxicity can be determined. Emission charges are calculated base on the quantity and toxicity of the emissions.   This method promotes and encourages technological innovation as there is an inducement to minimize pollution by use of better methods / technologies, inputs or even change of outputs. Secondly, the compliance costs are reduced, i.e. those who comply pay less charges e.g. they may have installed abatement structures that enable them to release non-polluting emissions. It is also a viable system for revenue for the government, and offers flexible options for polluters (Djikstra (2003), Canter (1977) and Lee and George (2000).

Generally, the marginal cost of pollution abatement is inversely proportional to pollution emissions.   The reverse is true for marginal damage cost. At a point where the marginal cost of pollution equals the marginal damage cost, there is a balance, and is the most efficient pollution level for a form. Anything less or more makes it easier to pay the charge, and anything beyond makes it cheaper to invest in pollution abatement technologies. It thus gives the polluters options that make economic sense, while it encouraged research and development (Dijkstra, 2003).

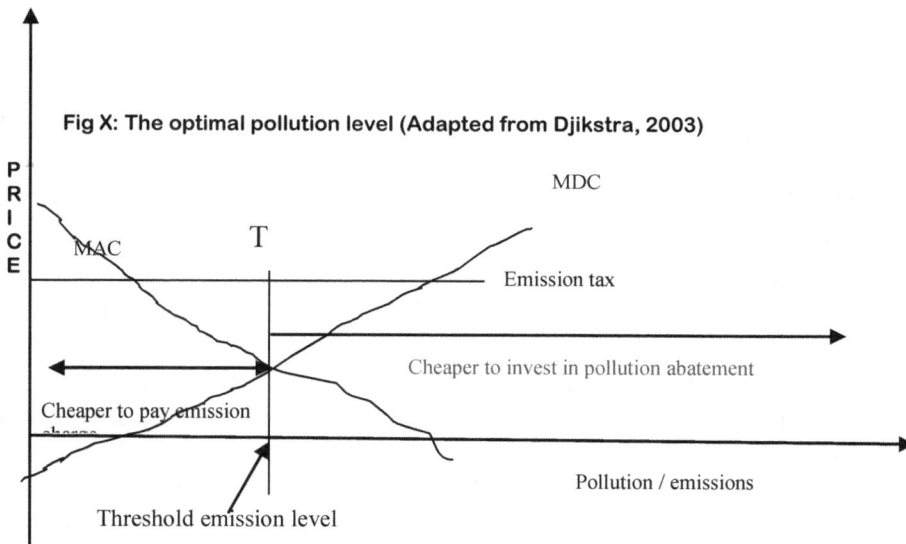

Fig X: The optimal pollution level (Adapted from Djikstra, 2003)

TRADABLE POLLUTION PERMITS (TPPS)

Tradable pollution permits are documents released by a government or its pollution control arm, and sold to those who need them to have rights to pollute. The number of these permits is properly marched with the threshold pollution level, so that one permit allows one to pollute only up to some maximum level. Thereafter, one has to get more rights (by purchasing another permit). The numbers released by the authority are set after determining the available threshold pollution level. If, after release of these permits, it is felt there is need for some to be shelved, the authority may buy some of the permits from the owners who bought them, and thereby match any pollution level with the environmental capacity to accommodate the pollutants (Kgathi and Bolaane, 2001). In a similar way, environmental activists may purchase and hoard some of the permits so that they are unavailable to whoever wants to use them to have pollution rights. The problem with this is that the extremists may hoard so much that industrial and related production processes may significantly decline, thereby introducing another dimension to the environmental management scene ((Dijkstra (2003); Lee and George (2000) and Canter, 1977)).

MORAL PERSUATION APPROACH

Moral persuasion approach in pollution control is a method of public involvement in pollution control (Kgathi and Bolaane, 2001). This may take many forms, but in its simplest form, it involves organizing training and related workshops for polluters. In these workshops and training sessions, they are given the necessary training skills required to help them appreciate the need for pollution control, as well as convince them to invest in pollution abatement systems and technologies.

In another form, the moral persuasion instrument may involve educating the public on a wide range of issues pertaining to environmental protection; they become custodians of the environment. This may be called environmental education. In such cases, the aim is to empower the public sufficiently to be able to put pressure on polluters to install abatement structures or close down. The belief is that the public, being the immediate stakeholders in whichever part of the environment and location of the polluter, will always be affected in one way or the other, and can cal for immediate action. Such may lead to participatory environmental management methods that may succeed to some extent (Kgathi and Bolaane (2001) and Afullo (2003)).

However, the public may be moved more by emotions than by the reasoning once they are aggrieved (Afullo, 2003). There is therefore the possibility to reason emotionally and by impulse, leading to even potentially useful projects being abandoned. This may be worse if the agenda is picked by credible rights and / or environmental activists. Thus whereas moral persuasion approach may be effective, and fast, it works best if combined with other methods, such as standards and economic instruments.

CONCLUSION
Generally, no single instrument can play the role of reducing environmental degradation alone. They have to be used in combination. Their effectiveness varies with circumstances. Thus whichever should be applied at a particular time needs thorough assessment. Use of social, economic, command and control / legal and related instruments all help in curbing degradation of our natural assets, thereby guiding us towards the broader objective of sustainable development. Tools such as strategic, cumulative, sectoral, and related forms of environmental assessment are therefore useful in deciding which instruments could be most effective in environmental protection. Lastly, before the choice of an instrument is done, proper consultation needs to be done to enlist public support; without incorporating this destructive yet essential subset in the environmental destruction and protection game play, nothing is

complete- for most of the time it is their long term interest that is paramount-whatever environmental jargon is used.

REFERENCES
1.   Afullo, A (2003). The Environment: Some concepts, issues and Concerns. Sustainable Futures (SF), Maun, Botswana. ISBN 99912-0-461-X
2.   Canter, L (1977) Environmental Impact assessment. 2nd edition. Mc-Graw Hill International Editions. Civil Engineering Series. ISBN 0-07-114103 0 . Singapore.
3.   Dijkstra T (2003. Environmental Assessment. A WEDC Post-graduate Module. WEDC, Loughborough University.
4.   Kgathi, D and Bolaane B (2001) Instruments for Sustainable Solid Waste Management in Botswana. Waste Manage Res 2001: 19: 342-353
5.   Lee N and George C (eds)(2000). Environmental Assessment In Developing And Transitional Countries. John Wiley and Sons Ltd. Sussex.

ANALYTICAL STRATEGIC ENVIRONMENTAL ASSESSMENT
Using a case study of an environmental impact assess met of your choice, reflect on the decision-making process using analytical strategic environmental assessment framework as a reference methodology.

Analytical Strategic Environmental Assessment (ASEA) is a modified and a more systematic form of the formal Environmental Impact Assessment (EIA), which strives to improve the decision-making aspect of the environmental assessment process (Dijkstra (2003), Lee and George (2000) and Canter (1977). EIA creates an impression that decision-making is at only one point / instant (i.e., instantaneous), and involves only one person / party, the regulatory authority. In real sense, however, environmental assessment involves many decisions, made by different people/parties, and at different stages. Some of these decision makers assumed by the EIA, and areas where they make decisions in the project cycle are:

•   The financier: can decide to fund or not to fund the project at any stage even after initial financial approval. An example is the Koro-Imbuse project in Kenya, which was recommended by almost all internal authorities, but stopped by the civil society at implementation stage. This created a chain reaction- where the financier stopped the process; the then central government also lost interest, etc.

•   The developer: may also change mind at any stage, thereby stopping all other subsequent stages. This may result from change of policy, priorities or governments.

•   The civil society: This has a very strong mobilizing power both locally and internationally, and can stop even projects approved by the highest authority in the land, or give the project a very bad publicity. An example is the case in Botswana Bushmen relocation from their customary homes. This was brought to the international limelight by an international NGO. Many residents refused to relocate (especially after the publicity) despite threats of forceful eviction, and only moved after essential facilities and services were completely withdrawn. Another example is the case of KEL chemicals in Thika, (Kenya) in which a priest mobilized faithfuls to demonstrate against pollution from the fertilizer and sulphuric acid manufacturing industry.

• The politicians: can decide that no such a project is developed in his / her constituency or ward, and may mobilize constituents for backing;

• The central Government: The president can give an executive directive (superior to the decision of the regulatory authority).

• Parliament: the parliamentary committee can also decide the fate of a project.

• Host community: Can demonstrate for or against a project on their own volition.

• International laws on commonly shared resources e.g. international resources such as waters (e.g. Okavango delta (Botswana/Namibia); East African Lake Victoria (Kenya/Uganda/Tanzania). In the case of Lake Victoria, a law bars any East African country from developing projects involving drawing significant amounts of water from the lake- because of an existing agreement between the Colonial East Africa and Egyptian Government in 1929 (Afullo, 2003).

• The professional body(ies): can decide the fate of a project;

• The Landowner: Can change mind about sold or leased land deals. At times willingness to accept compensation prevails so that despite the developer's lucrative compensation offers for resettlement of the resident(s), the latter may be adamant, thereby completely installing, or delaying a project. This may change financial plans, which may render the project uneconomical.

• The construction sector: can dictate terms on whether to support a project or not; they may sabotage a project (e.g. if tender offered unfairly) even if approved by the regulatory authority- leading to its collapse.

• The neighboring residents or country: An example is the proposed POPA falls dam for generation of hydroelectric power by the Namibia Government. Because the proposed dam intended to use the internationally shared waters. This proposed project has been vehemently rejected by the Botswana Government citing ecological repercussions. This is a situation where cumulative environmental impact assessment ought to have been conducted.

In general, decision-making can take one or a combination of the following forms (adapted from Dijkstra, (2003; Lee and George (2000) and Canter (1977):

Dictatorial: One dominant decision maker (e.g. regulatory authority) makes a unilateral decision without engaging the other stakeholders at all. This has been the most dominant system in most governments and internationally funded projects. The decision lacks credibility, and its product mostly unsustainable;

Participatory: Where the regulatory authority acts as a participant alongside other stakeholders in a process facilitated by a professional body. Reasoned participatory decisions are made in such gatherings by serious deliberations and consensus. After this stage, no deviations are expected from the final authority- only approving the collective decision reached by the gathering.

Professional approach: Where the regulatory authority delegates the decision making to an appointed body (e.g. a consultancy team), which them makes its recommendations- which are then applied as it is. The consultancy may involve the publics and stakeholder sessions. However, these gatherings are rarely sufficient. The regulatory authority comes in just as an observer during the process to avoid influencing the direction of decision to be reached by the professional body.

Combined / Hybrid approach: This is whereby the participatory process is reinforced with the professional approach. The suitability hybrid approach depends on the nature of the project. However, it is advantageous because sufficient technical aspects are availed (which is likely to have weight in decision making by the regulatory authority) and shared among the stakeholders. If the technical aspect is given a participatory face, then the decisions reached by hybrid approach are likely to be most appropriate. The need for ASEA arose due to the following limitations of the decision making stage of the EIA

a. EIA is a project – based assessment whose results are not applicable to a wide range of other situations. As such, it is expensive to carry out for every proposed development project.

b. There are many social, political and regulatory constraints in the formal EIA.

c. EIA never takes into account temporal and spatial aspects of other related projects, i.e. the cumulative effects are ignored, making the project under consideration seems like an island around which nothing else happens.

d. The decision makers may take the top-bottom approach (i.e. dictatorial) so that the other stakeholders are simply informed of the decision without prior adequate involvement.

e. Delays in decision making often lead to other stages of the project progressing, leading to less, if any, incorporation of any recommendations into the rest of the project cycle;

f. Decisions are never made open to other stakeholders, leading to reduce contribution. This compromises the quality and acceptability of the final product.

g. When decisions are made open / public, there are often cases where no explanation or reason is provided for the decisions, some of which never seem to have taken into account the inputs from other stakeholders; (Dijkstra, 2003)

ASEA therefore strives to help improve the decision-making stage(s) of the EIA by ensuring basic rules of are observed; these can be considered as the terms of reference (TOR) of the ASEA. ASEA thus operates on the following tenets (as adapted from Dijkstra T (2003); canter (1977); and Lee and George (2000)):

i. Timeliness: Decision made at the right time so that it can be used appropriately at the next stage in the project cycle

ii. Transparency / accountability: Decision made in an open, systematic way, which can be easily understood by the stakeholders; the decision makers are also ready to take responsibility of the consequences of the decision made.

iii. Inclusiveness / Wide participation: This means appropriate consultation and stakeholder / public involvement is ensured before any decision can be made. This gives the decision more acceptances, and the decision makers acquire some respect and credibility.

(v) Credibility: An open, accountable, systematic, inclusive and transparent decision-making process imparts virtues of credibility into the process and on the decision makers. This gives them more respect and trust. The final document and project resulting thereof is also credible and more acceptable to the stakeholders.

(vi)   Comprehensive: This means that the decision made should have taken into account all possible factors and inputs in the circumstances.

ASEA is a more systematic method which strives to combine the dual positive attributes of the formal EIA and the decision making sciences, as is hereby illustrated in the figure below.

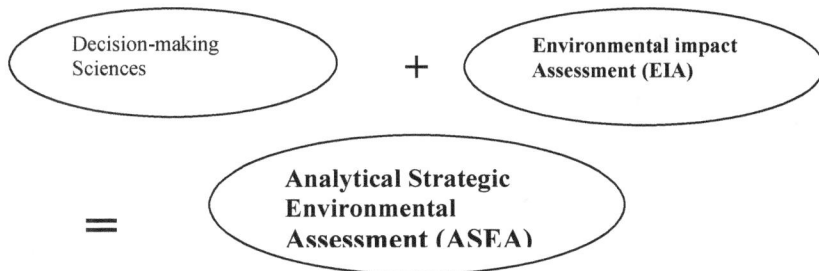

Decision-making Sciences **+** Environmental impact Assessment (EIA)

= Analytical Strategic Environmental Assessment (ASEA)

Adapted from Dijkstra T (2003)

ASEA, being a systematic procedure, goes through the following stages (as adopted from Dijkstra, 2003; lee and George, 2000; and Canter, 1977):

1.   Problem identification
2.   Gathering of information
3.   Public / stakeholder involvement / further generating and sharing of information
4.   Decision-making / implementation
5.   Monitoring
6.   Evaluation of the situation (a follow-up)

This systematized decision-making process makes it accountable, and based on the other stages of the cycle, including consideration of cumulative effects of the project. The decision making process is therefore treated as a continuum, taking place throughout. All these positive attributes render ASEA more credible, especially when combined with the strategic environmental assessment which goes beyond the narrow project-based approach to a wider and more open and sustainable policies, plans and programs. These two tools can be more applicable in the developing countries with no EIA legislation, and most projects pollute unsustainably. Incorporating the Strategic and analytical environmental aspects into third world (LDCs)development can also help reduce the cases of projects which have failed environmental standards elsewhere being imported into these countrie, as the LDCs desperately seek partnership with powerful countries in the name on welcoming investors.

BIBLIOGRAPHY
1.   Afullo A (2004) Environmental and Occupational health aspects of Waste Management in Maun, Okavango Delta, Botswana. An unpublished PhD Thesis, Commonwealth Open University, Asturias, Spain.
2.   Afullo A (2003) The Environment: Some Concepts, Issues and Concerns. Sustainable Futures, Maun, Botswana.
3.   Afullo A. (1995) Pollution of Lake Victoria by inorganic fertilizers used in the West Kano Rice irrigation Scheme. An Unpublished Mphil Thesis, Moi University, Eldoret, Kenya.
4.   Canter, L W (1977)(2nd edition). Environmental Impact Assessment. McGraw Hill International Editions: Civil Engineering Series. 2nd edition. Singapore. ISBN 0-07-114103-0
5.   CIEH (1998) Health and safety: First Principles. Chartered Institute of Environmental health, Chadwick House Group Limited, United Kingdom.
6.   Connell D W and Miller G J (1984) Chemistry and Ecotoxicology of pollution. John Wiley & sons, Brisbane, Australia.
7.   Dijkstra T (2003) Environmental Assessment. A WEDC Postgraduate module. WEDC Loughborough University, Leicestershire.
8.   GoB (2003). National Development Plan 9 (2003/4-2008/9) Ministry of Finance and development planning, Botswana Government.
9.   Lee N and George C (2000). Environmental Assessment in Developing and Transitional Countries. John Wiley and Sons, Manchester. ISBN 0-471-98557-0

CASE STUDY OF AN EIA

ENVIRONMENTAL IMPACT STATEMENT FOR THE PROPOSED BOREHOLE PROJECT FOR THE WANG' CHIENG' SUBSTATION, ODINO PROJECT, NDHUKLUHUNDA, TIANG, KENYA.

| Lead EIA Expert: Dr. Augustine Afullo, PhD (Lead Expert, NEMA Reg. No. 0468) P.O. BOX 5541-00200, CITY SQUARE, NAIROBI, Kenya Mobile: 0720640692/ 0734483934 afullochilo2002@yahoo.com / afullochilo@yahoo.com.sg | Proponent JJJJ CORPORATION, P.O. BOX 3315-40100, KISUMU Phone: 0572025637/8: |

PLANNING CONSULTANTS

NAME	QUALIFICATIONS	INSTITUTIO AFFILIATION
Dr Augustine T AFULLO, PhD. (Lead Expert, Reg. No. 0468), P.O. BOX 5541-00200, CITY SQUARE, NAIROBI, Kenya	PhD (Environmental & occupational health); Mphil (Environmental Studies); MSc (Water and Environmental Management / WEM) (Loughborough University @ WEDC); BSc (Agriculture); PGDE, PGDESc, and PGDHE.	ASEMBO INTERNATIONAL UNIVERSITY (KU), Department of Public Health, P.O. BOX 43844, NAIROBI, Kenya.
SUBMISSION OF DOCUMENTATION I, Dr. Augustine T. O. AFULLO, on behalf of Kinden Corporation, submit this Environmental Impact statement for the proposed Borehole at Koro-Imbuse Substation, Ndhukluhunda. To my knowledge all information contained in this report is accurate and a truthful representation of all findings as relating to the project. Signed at ....on this ....day of .... Signature.. Designation: EIA/Audit Lead Expert –NEMA Registration Number 0468		

| SUBMISSION OF DOCUMENTATION I, ...on behalf of the Kinden Corporation, submit this Environmental Impact Statement for the proposed Borehole at Ndhukluhunda. To my knowledge all information contained in this report is accurate and a truthful representation of all findings as relating to the project. Signed at ....on this ....day of .... Signature.................... ....Designation: .... |

Environmental Impact Assessment Guidelines

The environmental impact assessment (EIA) guidelines require that environmental impact assessment be conducted in accordance with the issues and general guidelines spelt out in the second and third schedules of the regulations. This includes coverage of the issues on schedule 2 (ecological, social, landscape, land use and water considerations) and general guidelines on schedule 3 (impacts and their sources, project details, national legislation, mitigation measures, a management plan and environmental auditing schedules and procedures).

Scope of the study:

This Terms of Reference (ToR) is developed by the Environmental Impact Assessment (EIA) consultant in consultation with the proponent, Kinden Corporation, for approval by NEMA (hereafter called the authority). The Environmental Management coordinating EMCA act, (1999) demands that an EIA be conducted in accordance with such a ToR. The goal of the EIA shall be to ensure that decisions on the proposed borehole project and activities are environmentally sustainable. The objectives of the EIA shall be:

i. To establish and document the baseline data at the proposed borehole site;
ii. To establish the proposed activities of the proposed borehole project;
iii. To predict the ecological impacts of borehole project at the Koro-Imbuse Substation, Chieng';
iv. To predict the health and safety impacts of the proposed borehole project;
v. To assess the potential socio-economic impacts attributed to proposed Borehole project;
vi. To identify mitigation measures for predicted socio-economic, health and safety, and ecological impacts due to proposed Borehole project;
vii. To propose alternative locations, scale and technologies to facilitate environmental management by the proposed Borehole project;
viii. To propose an environmental management plan (EMP) for the identified impacts alongside mitigation measures

The consultant shall undertake the EIA study to cover all aspects of the business, up to the recommendation and the Environmental Management Plan (EMP), in close collaboration with the proponent.

Responsibilities
While the environmental impact assessor shall provide the technical understanding on the baseline environmental status, projected impacts, impacts management, technological options and legal framework, as outlined in the output, the proponent was expected to provide the following:
• Site map(s) showing roads, service lines, buildings' layout and the actual size of the site;
• A valid Hydro geological survey report from a qualified hydrologist;
• Project quotation;
• Full proposal for the proposed project;
• Any clarification in cases of modifications and deviations from the written proposal;
• Indication of the preferred proposed location of project, as well as alternative sites,
• Full details of raw materials, process outline and anticipated by-products,
• Operation permits, registration certificates, licenses, approval letters,
• Land ownership documents and site history,
• Project budget outline (Bill of quantities);
• Meet all the costs of undertaking the study- including transport, communication, logistics;
• Provide key information on the proposal as may be required by consultant from time to time;
• Implement and adhere to the EIA report requirements and NEMA recommendations arising thereof.

Environmental Screening
This step was applied to determine whether an environmental impact assessment was required and what level of assessment was necessary. This was done in reference to requirements of the EMCA, 1999, and specifically the second schedule. Issues considered included the physical location, sensitive issues and nature of anticipated impacts. This helped inform the decision that a full EIA may not be necessary for this single borehole project.

Environmental Scoping
The scoping process helped to narrow down onto the most critical issues requiring attention during the assessment. Environmental issues were categorized into physical, natural/ecological and social, economic and cultural aspects.

Environmental Impact Assessment Output

The environmental impact assessment output from the consultant includes an EIA study report, which covers the following, among others, as per the EMCA, 1999 requirements:

i. Type of proposed activity;
ii. An indication of material and energy inputs, final products, by-products and wastes to be generated;
iii. A description of activities, processes and operations of the proposed project;
iv. Description of national legislative and regulatory frameworks on ecological and socio-economic matters relevant to the proposed project;
v. A description of the potentially affected ecological and socio-economic matters;
vi. A prioritization of all expected and / or on-going concerns;
vii. Identification of ongoing and / or expected environmental, occupational health and safety concerns;
viii. Alternative project options- locational / siting, size / magnitude and technological options in relation to the key predicted impacts;
ix. Recommendations for corrective activities, their costs, timetable & mechanism for implementation;
x. Measures taken to ensure implementation is of acceptable environmental standards; and;
xi. Non-technical summary outlining the key findings, conclusions and recommendations of the auditor.

Methodology Outline

This environmental Impact Assessment was carried out between 19th and 30th June 2006. The methods used included a guided tour of the site, Key informant interviews, Interviewing of workers at the proposed borehole site, Observation and listening survey, transect walk, desk top study (secondary literature), and photography to capture key features and views about the project.

Key Informant interviews **was extensively done with people with key information about the proposed development, those with area baseline information and those with technical information.**

Site visits and Assessments were made to Ndhukuluhunda Borehole project property and its environs. The visits were meant for physical inspections of the site and neighbourhood characteristics to determine the anticipated impacts. It also included further interviews with random members of the community. It helped shed light on current water sources, their advantages and disadvantages. **Rating and scoring, on the other hand,** was used to assess the different water source, site, and technological options to determine whether the Borehole project is necessary or not, locating / siting the Borehole, and deciding water source options. The scores, ranging from 1 to 5, were assigned per option, and the sum was determined, then the % calculated. The highest % was considered the best option, and vice versa. Finally, reporting was done as a requirement by EMCA. On basis of the above, coupled with constant briefing of the client, this environmental Impact statement (EIS) report was prepared. It comprises the general report, and a non-technical summary outlining the key findings, conclusions and recommendations of the assessor.

Other sections of the report are as shown below:

- Executive Summary
- Policy, Legal and Administrative Framework
- Description of the environment
- Description of the Proposed Project
- Significant Environmental Impacts
- Socio-economic analysis of Project Impacts
- Identification and Analysis of Alternatives
- Mitigation Action/Mitigation Management Plan /
- Monitoring Programme
- List of References including Appendices e.g. documents, photographs, unpublished data

The contents were presented as a report for submission to the authority as required by law.

ToR APPROVAL
STAKEHOLDER..................... Name ....
SIGN................DATE.....
Lead Expert / Consultant...
Proponent...
NEMA...

EXECUTIVE SUMMARY
Project type and Location:
This EIA report is for a single Borehole in Yawo Sub location, Soin Location in Tiang District. It is to be situated inside the Koro-Imbuse substation, Chieng', for domestic use by Kenya power and lighting company staff. The project was authorized on 15th May 2006, by the Lake Victoria South water services board to abstract 10m^3/day.

Methods

The methods used in this study included Key informant interviews, Environmental screening and scoping, desktop studies, interviews, site inspection, photography and reporting as documented in this EIA Report.

Expected activities:
Kinden Corporation expects to undertake the following Borehole related activities:

STAGE I: DRILLING WORKS using Rotary drilled down the hole (DTH) method, will do the following:
• Mobilization and demobilization of the drilling unit;
• Erecting (dismantling of drilling unit);
• Drilling 203 mm diameter borehole from 0-100m;
• Drilling from 100-200m (A target of 120 m predicted by the hydrological survey);
• Supply and installation of 152mm diameter plain black steel casing class B;
• Supply and install 152 mm diameter machine cut black steel screens class B;
• Supply and install 152 mm diameter machine cut uPVC screens class B;
• Supply and install filter gravel pack; and Allow standby time;
• Allow for supply and installation of surface casings on request by borehole drilling consultant.

STAGE II: DEVELOPMENT WORKS will involve the following:
• Test pumping;
• Construction of a mass concrete plinth around well head of 1.5 x 1.5 x 0.5 m;
• Borehole capping;
• Water chemical analysis and borehole completion report.

Project cost:
The project is to cost Kshs 515,000, covering site preparation, drilling and development.

Impacts:
▪ The proposed site is located within an environmentally insensitive area;
▪ The area lacks enough potable water facilities; the borehole water is likely to cause enamel fluorosis;
▪ Site has no unique fauna, flora, water, mineral or soil resources different from the surrounding;

- Some health and safety issues are inevitable, such as noise at drilling and development stage;
- The activity has no effect whatsoever on ecology and existing human settlement trends.
- Social and economic activities of the community are likely to benefit,
- There are no similar water projects in the vicinity.

Conclusions:
The Ndhukluhunda borehole project is technically feasible and ecologically sustainable, with no significant environmental impacts. The only expected environmental impacts are risks of enamel fluorosis, sinkholes and subsidence risks. The borehole drilling should be approved by the various authorities.

Recommendations
- A drilling permit be sought from the relevant board;
- A 2-litre water sample from the borehole be fully analysed before use to determine suitability for drinking;
- A monitoring tube be installed in the borehole to enable monitoring of water level;
- A master meter be installed at the borehole to monitor groundwater level;
  - Drilling and test pumping of the borehole be supervised by a registered hydrogeologist.
  - Subject to water quality results, defluoridation should be done to make the water wholesome and healthy.
  - The proposed Borehole should be allowed, and given an EIA license without delay, as it is likely to have more positive impacts than the adverse ones, subject to incorporation of the proposals shown in the environmental management plan.

Another diagram that could be used as a basis for this discussion is shown below.

Level of government	Land-use plans (SEA)	Category of action and type of assessm Sectoral and multi-sectoral actions		
		Policies (SEA)	Plans (SEA)	Programmes (SEA)
National / federal	National land-use plan	→	→	→
Regional / state	Regional land-use plan			
	Sub-regional land-use plan			
Local	Local land-use plan			

B+

b) **Construct a flow diagram illustrating a generic methodology for health impact prediction and assessment. Describe the main elements and the linkages shown in your diagram in no more than 400 words.**

This is a good answer, with the main elements are well discussed. Another possible solution is the one discussed in Canter on page 535. The diagram of a generic methodology for health impact prediction and assessment is illustrated below.

B+

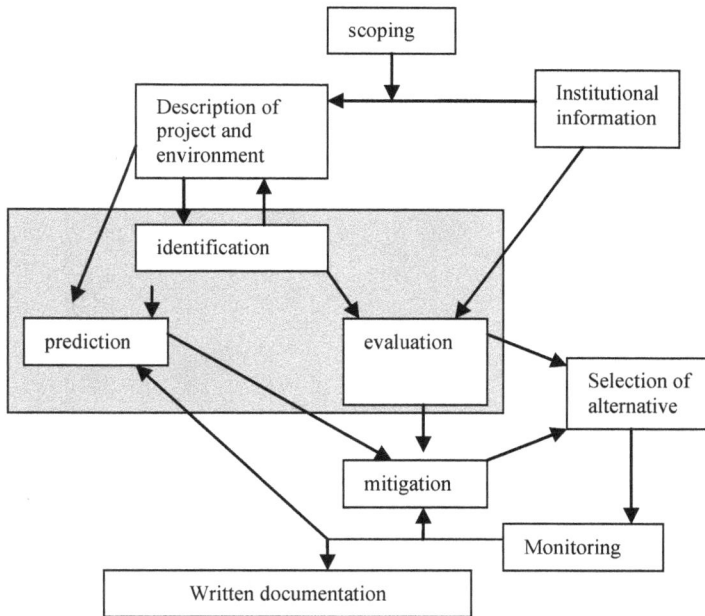

CRITIQUE OF THE ABOVE MODEL
A case study should illustrate the following:
• what are the manifestations of environmental degradation (overuse, waste, inefficiency), low rate of return of environmental assets, unsustainability, mono rather

91

than multi use of environmental assets, lack of resource protection and environmentally enhancing investments, diminishing returns, lack of access to improve stewardship, consequences of exploitation of environmental assets not taken into account in environmental decision making, low recycling rates and disposal costs not calculated, biodiversity loss and pollution

- process of environmental degradation, policies, plans and programmes that are relevant to addressing degradation;
- effectiveness of policies, plans and programmes;
- economic aspects (relevance of OARs and CPRs) and relevant opportunities to changes attitudes towards environmental degradation - this discussion should address issues such as universality, exclusivity, transferability and enforceability
- examples to link to could include the tragedy of the commons case;
- it should address national and international contexts

Question 2a

Should have included issues related to the following points:

- Human perception is not constant since it is influenced by levels of education, cultural background and availability of information. It influences the WTP for global environmental change which may vary within a community and dictated by human perception of direct benefits to human, productivity, health and indirect benefits.
- WTP and WTA vary from country to country since they depend on the priorities of individual countries. Low income countries tend to give more priority to economic benefits (such as production, manufacturing) over the environmental quality.
- WTP and WTA may vary according to the perception of risk across the community. Risk perception contributes to a community's WTA changes in their behaviour in order to avert dangers.

Question 2b

The question required the use of a case study of an EIA and to reflect on the decision making process using the analytical strategic environmental assessment (ASEA) framework. This response has one through details of ASEA and some of the issues related to ASEA. It really should have related what is presented as a case study identifying strengths and weaknesses and indicating options for improvement.

SECTION 3: WASTE WATER TREATMENT

WWT TOPIC TASKS:
BACTERIA AND WASTE TREATMENT

Question1. Explain which groups of bacteria are of greatest importance for aerobic self-purification and waste water treatment, and the roles that these groups perform.

There are many kinds of microorganisms, especially bacteria, playing the role of aerobic decomposition of organic and inorganic wastes in effluents. Aerobic self purification bacteria operate in an aerated environment, i.e. one in which molecular oxygen is readily available, and acts as a reactant in the decomposition process. These groups are given in the table below:

Bacteria group	Role played	Example
photoautotroph	Use light to photosynthe, releasing oxygen which facilitate further substrate decomposition	Photosynthetic bacteria
Photo Heterotrophic bacteria	Break down organic carbon from effluent as source of their energy for growth and multiplication. This reduces BOD loading of effluent.	
	Reduce the inorganic loading of effluent by taking up components of the inorganic constituents	Some sulphur bacteria
Nitrifying bacteria	Convert ammonia gas to nitrites; Convert nitrite to nitrate; **BOTH** Reduce the ammonia loads in the waste, thereby making it less toxic for other organisms, as well as reduce the alkaline pH associated with ammonia.	Nitrosomonas Nitrobacter
Chemorganotrophs	Derive their carbon from the organic compounds in the effluent, thereby reducing the BOD.	
Chemolithotrophs	Derive their energy from oxidation of reduced organic compounds (REDOX reactions	

OTHER MICRO ORGANISMS IN WASTE WATER TREATMENT

Essay: Other microorganisms other than bacteria of importance to self-purification and wastewater treatment.

Besides Bacteria, many other classes of micro-organisms are involved in effluent treatment. They play complementary roles, ending up with refined, stable and settled wastes. These micro-organisms are protozoa, viruses, algae, fungi and rotifers. They are discussed in the table below.

Organism	Importance in effluent treatment and purification
Fungi	Degrade cellulose and other difficult organic matter e.g. polysaccharides through the hydrolytic process; this facilitates further breakdown of these inert organic matter; Useful in composting sludge.
Algae	Photosynthesize, thus increasing the concentration of Dissolved Oxygen (O_2) in the water / wastewater. This supports and facilitates other aerobic processes and organisms.
Protozoa	Feed on suspended particulate organic matter (including bacteria) as an energy source, thereby facilitating cleaning, clearing and polishing of the effluent treatment media. Dead and old bacteria when eaten rejuvenate the system e.g. in activated sludge and trickling filter systems. This feeding also helps balance (reduce) the number of micro-organisms and thus help balance Food to microorganism ratio (F/M ratio)
Rotifers	Consume dispersed and flocculated bacteria, and organic matter; indicator of efficient biological purification process
Viruses	Bacteriophages may reduce effectiveness of bacteria by incapacitating them- e.g. by inflicting disease. This may control the bacteria concentration, and therefore F/M ratio in established systems. This, however, may work negatively for systems where bacteria numbers are low, e.g. a new system under development.

Conditions necessary for waste water oxidation processes
Conditions necessary for bacteria for bio-oxidation to thrive and multiply.

Biological oxidation processes take place under the mediation of microorganisms. These are sensitive to stimuli, and have specific requirements without which they are completely incapacitated, or their activities significantly reduced. These requirements are nutrients (source of carbon), micronutrients, oxygen, appropriate temperature, suitable degree of acidity and alkalinity (pH) etc. These conditions are presented below.

Condition / stimulus	Optimum	Effect
Temperature	Mesophilic range (best at within 32 – 42°C)	Every 10 ^{o}C change double or halves the reaction rate below the optimum temperature for each species. High temperatures reduce dissolved oxygen level in water, thereby suppressing aerobic microorganisms.
pH	6.5 – 7.5	Extremes affects (mostly reduce) reaction rates; may immobilize some nutrients; each microorganism has its optimum pH, but works within a narrow range, and become incapacitated at extremes.
Micronutrients	Molybdenum, zinc, Selenium, cobalt, copper, etc	Catalyze reactions; essential nutrients which facilitate proper metabolism of other foods e.g. carbohydrates, and therefore facilitate BOD reduction and nitrification processes;
Major inorganic elements	Magnesium, calcium, sulphur, iron, Sodium, Chloride, phosphorous etc	Form parts of cells / cell constituents; growth; Major chemical energy carriers (e.g. ATP in case of phosphorous) etc
Oxygen	Reactant in the catabolic processes e.g. respiration / breakdown of carbohydrates for energy release	Obligate aerobes are killed by lack of molecular oxygen; facilitate aerobic reactions. Concentration also a good indicator of appropriateness of design.
Carbon source	Organic matter in effluent e.g. carbohydrates	For cellular growth and multiplication. The major nutrients such as carbohydrates, cellulose etc are hydrolysed

Calculation.

The inlet to a wastewater treatment works contains 3 equal constant-velocity grit channels. The cross-section of each grit channel approximates to a parabola of maximum width 0.8 metres and maximum depth of 1.0 metres.

a. Estimate the maximum possible flow through this installation. State any assumptions that you make.

Assumptions:

1: velocity of flow (V_f) = 0.3 m/s to allow grit to settle; should the velocity of flow be less than 0.2 m/s, organic solids may settle out; if it is more than 0.4 m/s, some grit may be carried forward.(WEDC, 2003; Crites and Tchobachnoglous, 1998))

Area of parabola (A) = $2/3$ bh (where b= width of grit channel; h = depth of grit channel)
If width = 0.8 metres; maximum depth = 1.0 metres;
Therefore A= $2/3$ (0.8 m) x (1.0 m) = $1.6/3$ m²
Flow through grit channel (Q_g) = V_fA (Where V_f means 0.3 m/s and A = area of parabola)
Therefore Q_g = 0.3m/s x $1.6/3$ m² = 0.16 m³/s
Therefore maximum possible flow through the constant velocity grit chambers is 0.16m³/s

Question 2b.   Estimate the approximate length of each grit channel.

Assumption: design velocity = 0.3 m/s; at this velocity, the grit will settle at 0.03m/s.
Therefore the theoretical length of channel required =
Depth of flow in channel (m) x velocity (V_f)/ 0.03   (NB: V_f = 0.3m/s)
Length = 10 x depth of flow in channel = 0.3/ 0.03 x 1m = 10 x 1 m = 10 metres.
However, from experience, longer grit channel length is required to allow grit of different sizes to settle and to compensate for turbulence.
This adjustment therefore comes to 20 x maximum depth of flow = 20 x 1 metre = 20 metres.
Therefore the approximate length of each grit channel is 20 metres.

Question 2c.

Explain how settlement of particles in a constant velocity grit channel differs from that in a primary sedimentation tank.

Grit particle concentration is generally low in effluents, thus they will behave as discrete particles. In this case therefore they will settle by unhindered settling process inn which they obey stroke's law in their behaviors while settling.

Strokes law states that the maximum settling velocity for a spherical particle in a liquid
(V_{Max}) is given by:
V_{Max} = $2/9$ (r²g/n)(p-p')   Where p= density of particle;
P' = density of liquid;
N = liquid viscosity;
G= acceleration due to gravity
r = radius of the settling particle.

Thus the maximum settling velocity of any grit particle is directly proportional to the difference between the density of the particle and that of the liquid, square of the particle radius, and the acceleration due to gravity. The velocity, however, is inversely proportional to the liquid viscosity. On the other hand, the primary sedimentation tank exhibits four kinds of settling. These are:

- Hindered or zonal settling; Due to high particle concentration, particles settle as a mass, producing a distinct liquid / solid interphase. Interparticle forces bind together and disallow independent settling.
- Unhindered settling (discrete –particle settling); as per strokes law described above.
- Flocculant / Coalescence settling; small particles combine to form larger ones which settle faster, sometimes as flocs;
- And compressive settling; occurs at tank bottom where all particles are in contact with one another; Thus further settling can only occur by water being squeezed out to allow sludge consolidation; the sludge particles realign, get squeezed, and forms sludge blanket in which particles are like in layers but closely packed.

Thus the main difference in sedimentation in grit channels and primary sedimentation tank is simply that the grit settle in more or less a uniform manner as discrete particles whereas the solids in a sedimentation tank, by virtue of being so mixed up and of many different sizes and shapes, behave like grit where the particles are heavy and few, to compressive pattern.

Essay: Suggest suitable depths and diameters for an installation of 3 circular primary sedimentation tanks which share equally the flow for waste water having typical properties as shown in Table Q3.

Assume suitable design values for important design parameters, such as the retention time and the surface loading. Your design need not be detailed, and for this exercise treat sedimentation tanks as being cylindrical, with vertical side walls and a flat floor. Explain how the simple cylindrical shape assumed differs from the shape of circular sedimentation tanks likely to be used in practice.

Dry weather flow (dwf)	8600 m3/day
3Xdwf	24650 m3/day
BOD (Raw sewage)	4320 mg/l
BOD (Settled sewage)	275 mg/l
SS (Raw sewage)	575 mg/l
SS (Settled sewage)	310 mg/l

Given BOD of raw and settled sewage, the % BOD reduction is (430-275) / 430 = 36%
This is within the expected treatment efficiency range of 25-40%.
AND
SS removal given raw and settled sewage SS levels; the % SS reduction is: 575-310)/ 575 = 46%. This SS reduction is a little below the normal expected range of 50-70%.
 And given
%BOD reduction = t/a+bt ....(i) (Where a=0.018 and b = 0.02)   AND
% Reduction in SS = t/a+bt ....(ii) (Where a = 0.0075 and b = 0.014)
Therefore
$36 = t/0.018 + 0.02 t$........................(iii)   AND
$46 = t / (0.075 + 0.014t)$........................(iv)
Solving equations (iii) and (iv) simultaneously, $T = 3.986$ hours $\approx$ 4 hours.
Therefore the retention time = 4 hours..(1)

This is in accordance with Crites and Tchobagnoglous (1998) and WEDC (2003) which state that for an upward flow tank, retention time can be at least 2-3 hours. This is appropriate because C and T, 1998 state that Design for retention times may be for peak flows when retention time is 1.5-2.5 hours; or for dry weather flows when the retention time may be between 6 and 8 hours. This, however, can be reduced in hot climates to reduce the possibility of septicity of the effluent. Thus 4 hours is acceptable.

According to C and T (1998), surface-loading range of at most 43 m^3/m^2d is typical, with a common average value of 29 being universally accepted. In radial flow tanks, a maximum surface loading of at most 45 is acceptable.
Thus this design will assume a surface loading (SOR) of $29m^3/m^2d$.................(2)

Using the surface loading (SOR) above to calculate the diameter of the tank, and
Using r to represent radius of tank, A to represent cross-sectional surface area of tank and h to represent depth,
Peak flow rate (3 x dwf) = 24650 m^3/d= SOR / A
$24650 = 29 \times 22/7 \times r^2$
Therefore $r^2 = {}^{24650 \times 7}/{}_{29 \times 22} = 270.45$ Thus r = 16.44 $\approx$ 16.5 m
Thus the diameter of the tank is 16.5 x 2 = 33 metres.....................................(3)

According to WEDC (2003) and Crites and Tchobagnoglous (1998), diameter of a circular primary sedimentation tank should not exceed 45 metres.

Lastly,
Depth of the tank (h):
Capacity of tank = $22/7 \times r^2 \times h = 8600$
H = $8600 / 22/7 \times (16.5)^2$
Thus h = 8600 / 855.41 = 10.05 m
Thus the depth of the tank is 10 m...(4)

According to WEDC (2003) and Crites and Tchobagnoglous (1998), the depth should be at least 1.5 metres.

The key difference between this design and the conventional one is simply in shape. The conventional type would be easier to collect sludge in whereas this one will have a separate means of sludge collection due to the flat bottom. It can therefore substantially reduce the depth of the tank, and may lead to reduced sludge collection and settling of suspended solids due to continuous turbulence which shakes the entire tank bottom. It thus can only work for calculation purposes to help conceptualize the whole process, but it is not practical in effluent management. However, it is deep enough to allow for this extra turbulence, as the forces are least likely to interfere with settled sludge at the bottom of this very deep tank.

ENVIRONMENTAL STANDARDS
In a region where effluent quality standards are to be introduced for the first time, discuss the relative advantages and disadvantages of setting:
(iii)   national emission standards (The uniform effluent approach); and
(iv)   Ambient standards for individual works (the water quality objectives approach) based upon the mass balance equation

The broad objective of regulatory instruments in pollution control is protection of receiving environments and public health. This is to ensure that effluents do not cause acute or chronic damage to the biotic and abiotic forces. Regulatory instruments involve deciding and enforcing measurable and achievable effluent standards. If well constituted, standards can be used to determine effluent fees, therefore acting as a way of sustaining, running (operation and maintenance of) and expanding the facilities. However, in systems where they are new, a lot of care need to be taken, as the whole process may cause unnecessary tension if the set standards, using whichever method, are neither enforceable nor achievable.

The water / effluent quality assessment instrument should have the capacity to measure the basic standards parameters, and interpret them accordingly with the specific setting in mind. The use of standards is generally referred to as the command and control strategy. It is generally known, from experience, to be cumbersome where there are no facilities and institutions to support, develop and enforce. An incremental command and control approach would be a compromise strategy for poor developing countries whose immediate priority is poverty alleviation, employment, and food production (all of which demand investment, sometimes even polluting facilities). Setting very stringent standards may in such cases deter development (industrialization and commercial production processes); such countries may not consider environmental conservation a priority, since the level of pollution is still negligible compared to the gains associated with the pollution facilities.

Two alternative approaches for meeting effluent quality standards are possible. These are:
- Ambient standards (Water quality objective)
- Emission standards (Uniform effluent approach).

The following table gives a comparison between these two approaches:

	Advantages	Disadvantages
Ambient standards approach	Focuses on the Water quality as an objective Suitable to the local environment; Takes into account local factors; Flexible and therefore more appropriate; Less capital cost; Varies with flow and final water use; Case-specific and therefore more appropriate in protecting the ecosystem; More participatory as decisions made more often; Timely self cleansing by receiving waters possible; Takes into account use of the receiving water body;	Difficult to measure correctly; Amenable to abuse; Needs regular field assessment to justify change of standards; Amenable to controversies; Can be improved on incremental basis; Does not consider water use;
Emissions standards approach	Focuses on uniform effluent approach Effluent of same standard; Less discriminative; Considers actual and potential water use; Uniform quality monitoring facilities; Easy to monitor and evaluate;	Difficult to enforce uniformity; Less participatory/ more alienating; Heavier capital outlay; Standards may be inappropriate for some situations e.g. during low flow seasons and times of the day e.g. nights); Needs a lot of baseline data; Likely to cause serious ecological problems; Difficult to agree on common standards to use;

THE OXYGEN DEMAND

The parameters BOD, COD and TOC (Biochemical oxygen demand, Chemical Oxygen demand and Total organic carbon) may all be used to assess the concentrations of organic pollutants present in wastewaters. Explain why it is not convenient to use just one of these parameters, and whether it is possible to make direct comparisons between these three different parameters. For a sample of raw domestic wastewater which parameter (BOD, COD or TOC) is likely to have the greatest value, and which parameter will have the smallest value?

Water and wastewater normally contain materials, inorganic and organic, which decompose. In the process, these materials need and consume O_2. Since the reaction takes place in water, the O_2 required is acquired from the water body. These materials are therefore said to create an oxygen demand (OD) on the water. The OD can be Biochemical (BOD) or Chemical (COD).

The Biochemical Oxygen Demand (BOD) can be measured in 2 ways: BOD_5 and BOD ultimate (BOD_u). Generally, BOD_5 is much smaller than BODu because the former is a measure of Oxygen consumed in only 5 days at a temperature of 20^0C. Within these days, only carbonaceous oxygen demand is catered for. If left alone to continue reacting after day 5, BOD would increase, as nitrogenous degradation by oxidation starts after week 1 of the BOD test. This is the main limitation of BOD_5; it only caters for the fast decomposed organic matter (largely C-based), even though it is a parameter commonly used to measure organic loading of water or effluent. Not even all carbon is taken into account, a gap normally filled by the carbonaceous Oxygen Demand. However, it is a quick measurement, and results are available in less than a week. It therefore does not involve an indefinite use of fuel to keep the oxidation reactions going.

BODu, on the other hand, is a more complete measure of the OD, as it takes much longer, giving a chance for all key organic but potentially oxygen demanding components to be oxidized. Thus, the more inert nitrogen-based compounds, which only get oxidized (through the nitrification process) from week 2 of the reactions, get a chance. Similarly, the less reactive C-compounds which could not be oxidized within the 5 days of BOD_5 determination get time to react. It thus gives a more realistic OD. BODu determination also occurs at normal ambient temperatures. This renders it cheap, as less energy is used, and reflects more closely the real OD as microorganisms use the organic C and N. it thus fully covers both the BOD_5 figures as well as the TOC values. The use of Chemical Oxygen demand (COD) alongside all these can be even more useful, as COD takes into account other inorganic oxygen demanding water constituents, as well as being short, requiring only 2-3 hours to complete. It is therefore normally the highest and the second fastest after TOC. TOC, on the other hand, may take 5-10 minutes to determine, and thus can be used in process control.

Thus it is clear that each of the three tests, BOD_5, BODu, COD and TOC have different values with meanings. Each has clear advantages and disadvantages. Yet having all of them demand time, skilled personnel, money and facilities. Thus whereas the best thing could be to have all the figures so that the gap left by one can be filled by the other(s) at interpretation time, it may not be practical. On the whole, the three parameters are related as shown below.

Generally, BOD_5 < BOD ult < COD, TOC
TOC is higher than COD if wastes are either domestic and / or industrial but organic e.g. from food processing, i.e. the organic matter is higher; COD will be higher if the effluent is largely inorganic in origin e.g. mining or metal refining industry.

For untreated domestic wastes, generally:
$COD \approx 2BOD_5$; BOD ult $\approx 1.5BOD_5$; and $BOD_5 \approx 2$ TOC;
This can be re-written as:
TOC: BOD_5 : BODult : COD $\approx\approx$ 1:2:3:4 (Crites and Tchobachnoglous, 1998)

These relationships, however change as the effluent undergoes treatment.For a sample of raw domestic wastewater, TOC will have the smallest value and COD the largest value. Going by the relationships above, one OD parameter from among the above can be determined accurately, and the value used to estimate the rest. This may not only make the determination cheaper, but also make interpretation more complete, and intervention more appropriate. Where there is a decision to be made from the values, it can be made more appropriately from a strong information base when all the parameters have been determined.

REFERENCING
1. Crites R and Tchobachnoglous G (1998). Small and decentralized wastewater management systems. McGraw Hill International editions. Singapore
2. Mara D and Pearson, H (1988). Design manual for Waste water stabilization ponds in Mediterranean countries. European Investment bank and

Mediterranean Environmental Technical assistance Program, in collaboration of Lagoon Technology International, leeds, England
3. WEDC (2003) Wastewater treatment: A post-graduate module. Loughborough, UK.
4. WEDC (2003) Wastewater treatment. Additional resources. Loughborough, UK.

CRITIQUE OF RESPONSES:

	Comments
Presentation	Good and clear. Use of headings and sub-headings would help to show the structure of your answers more clearly. We noted that you were unable to benefit from the message sent to participants in answer to some questions.
Question 1	The use of a table in part a was good, and your interpretation of the question was valid, although I was expecting more general descriptions of the characteristics of the most important bacteria. See the comments sent to all students. Part b was good. Ciliates are an important group of protozoa, and higher animals (worms, flies, etc.) may graze on other organisms. Part c was also good. See the comments sent to all students, which mention salinity, absence of toxic materials and a moist environment.
Question 2.	Flow rate for one channel is 0.16 m³/sec. Include more assumptions. Generally good answers.
Question 3.	Include references for the equations for BOD and SS reduction giving t (or T) = 3.986 hours. Your approach was correct, but you designed only one tank to treat the entire flow of 24650 m³/day, instead of sharing the flow between three tanks.
Question 4.	A good and considered answer to the question, considering practical aspects and the risks of standards not being enforced. The table summarised your points well. See the comments sent to all students.
Question 5.	Good discussion of the differences between the three parameters. A clearer comparison of what each parameter measures would be helpful. For more information see the comments sent to all students.
References	Good. Try to find additional references, and include references in addition to those provided with this module.

WASTEWATER TREATMENT DISTANCE LEARNING MODULE
TOPIC TASK 2
Alternative approach No 1.

	Domestic	Industrial	Combined	
BOD	275	925	418	100
Nitrogen	32	16	28.5	6.8
Phosphorus	10	4	8.7	2.1
Percentage	78%	22%		

The combined wastewater contains excess N (6.8 > 5.0) and P (2.1 > 1.0).
Question No 2.
(0.78 × 275) + (0.22 × 925) = 418
(0.78 × 32) + (0.22 × 16) = 28.5
(0.78 × 10) + (0.22 × 4) = 8.7
The ratio BOD/N/P = 418/28.5/8.7 = 100/6.8/2.1

b. Daily total water used = 160 × 140 litres = 22400 litres = 22.4 m³
Daily total wastewater = 0.8 × 22.4 m³/day = 17.92 m³/day
BOD load = 440 × 17.92 grams/day = 7884.8 grams/day
Area of discs needed for hydraulic loading = 17.92/0.035 = 512 m²
Area of discs needed for BOD loading = 7884.8/6.5 = 1213 m²
Area of 150 discs each 2.0 m diameter (both sides) = 150 × 2² × π × 2/4 = 942 m²

Area of discs needed for hydraulic loading = 512 m²
Area of discs needed for BOD loading = 1213 m²
Area of discs provided = 942 m²
The area of discs appears to be insufficient, but:

- The BOD reaching the discs will be reduced during primary sedimentation.
- The area of discs will be greater than calculated because discs are not flat but are corrugated.
- On the other hand, the full disc are cannot be used. (A small area near the axle is not used.)

Question No 3.
Submissions were very good, with good designs. The standard of submissions was very good. Some of the general comments are:

1. Choose appropriate length to width ratios for some ponds. See section 7.5 of the Design manual for waste stabilization ponds in Mediterranean countries.
2. Ponds can have more than one inlet and outlet to help distribute flows and avoid concentrated flows.
3. It may be advisable to share the flow between two parallel sequences of ponds.
4. It is helpful to provide a summary table showing the total land area needed for the pond installation, the total retention time for the series of ponds, etc.
5. What crest width is desirable on embankments between ponds to permit access?
6. Some people commented on the required effluent standards. Effluent standards, and parameters, need to be chosen to suit local circumstances (river quality, possible downstream use, etc.). Additional parameters could have been considered. The algal content of the effluent may be important, because discharge of algae to receiving waters needs to be controlled. Calculations for reduction of total and ammoniacal nitrogen, as shown in the design manual for Mediterranean Countries, are inconsistent. They suggest that the ammoniacal nitrogen content in the effluent is greater than the total nitrogen content.
7. The inlet to the series of ponds should be on the town side of the installation. Flow through the ponds should be towards the South-west, away from the town and in the direction in which the ground slopes down. There is no need to assume a minimum head difference between adjacent ponds: there is likely to be a

difference in water levels if the site is on a slope, and water levels will be dictated by the level of the outlet structures. Note that ponds need not have a uniform top water level.

8.   Ponds need not be arranged in a long straight line, but can be arranged in a more compact arrangement.

9.   It is possible to calculate the length of outlet weir necessary. Various designs for inter-pond connections are possible.

10.   Ponds in parallel but not in series. It is expected that under normal conditions the majority of the water in a facultative pond will contain an appreciable quantity of dissolved oxygen and that, as a result, the biological stabilisation processes will be largely aerobic. It is therefore essential that the BOD of the pond water is not excessive. Normally it is considered that the BOD of a well-mixed facultative pond should not exceed 60 mg/l although there may be little harm in it exceeding this figure by even as much as 20 mg/l. It is also largely for this reason that facultative ponds are not built in series. If the BOD of the pond is appreciably in excess of 60 mg/l then it will no longer be facultative but anaerobic, and if there is a pond in series following a similar pond producing a 60 mg/l BOD effluent then it will nearly certainly be fully aerobic, (i.e. a maturation pond). It is therefore not possible to have two facultative ponds in series.

11.   A contents list would have been helpful for some reports.

12.   Most people realised that the first maturation pond needs to be designed for BOD loading (see section 6.5.1 of the Design manual for waste stabilization ponds in Mediterranean countries.

13.   It is desirable, but not essential, to provide screening and grit removal facilities before the pond inlet, so that easily-removed materials are taken out before they enter the ponds.

14.   Information about soil conditions and depth to water table was not provided, so the need for pond linings could not be determined.

15.   The surface loading design approach for facultative ponds usually gives a larger surface area than the rational approach, so I tend to use the surface loading method.

16.   Most people chose to provide the following series of ponds: Anaerobic, Facultative, First Maturation Pond, Second Maturation Pond. A few people provided an additional (third) maturation pond for bacterial removal, but two maturation ponds could achieve the stated effluent quality by adjusting the retention times of some ponds. For example, the surface loading approach provides the area required for a facultative pond, but the depth is not fixed. Increasing the pond depth from 1.5 to 1.75 metres increases the pond volume, and retention time, by about 16%.

17.   There is no single right answer to this question. I assumed embankment slopes of 1:3, 85% reduction in BOD in the Anaerobic and Facultative ponds, and 25% reduction in BOD within each Maturation pond. For a system with all flows split between two

parallel sets of ponds so that each set of ponds treats half the flow I obtained the following results.

Total area of ponds at mid-depth = 3.50 Ha
Total area of ponds at crest level, including freeboard = 4.28 Ha
Total retention time = 17.56 days

Type of pond	No. of ponds in parallel	BOD (in)	BOD (out)	FC or TTC (in)	FC or TTC (out)	Depth (m)	Length at mid-depth (m)	Width at mid-depth (m)	Retention time (days)	Freeboard (m)
Anaerobic	2	450	162	9×10^6	1.46×10^6	4.5	30.0	15.0	1.41	0.5
Facultative	2	162	67.5	1.46×10^6	4.60×10^4	1.5	200.1	40.0	8.34	0.5
Maturation (1)	2	67.5	50.6	4.60×10^4	2.54×10^3	1.5	149.1	29.83	4.63	0.5
Maturation (2)	2	50.6	38.0	2.54×10^3	$2.0 \times 10^2 = 200$	1.5	123.6	24.73	3.18	0.5

Note: "FC or TTC" refers to "Faecal Coliforms or Thermotolerant Coliforms".

18. These calculations do not take sludge accumulation into account. Sludge is likely to accumulate at 0.04 m³/person.year. For a population of 24000 people, the sludge accumulation in one year is likely to be 960 m³. This is a significant volume, when the volumes of the anaerobic ponds are (in total) 4320 m³.

CRITIQUE OF RESPONSES ABOVE

	Marks	Comments
Question 1	12.5%	Good. Clearly presented concise calculations.
Question 2	12.5%	Your approach was correct, and you made some good comments. However, each disc has two sides, and the BOD will be reduced by sedimentation before the wastewater reaches the discs. See the outline answer.
Question 3 Presentation	10%	Generally good and clear. Figure 2 was somewhat confusing because the arrows both identified features and showed flow directions.
Question 3 Structure	10%	Good and clear. A contents list, and some more headings and sub-headings could have shown the structure more clearly.
Question 3 Content	25%	Good explanations. I was not sure of the source for your assumed nitrogen and ammonia reduction percentages. There was a mistake in your final table, which showed 234000 FC/100 ml in the final effluent.
Question 3 Calculations	25%	I note that you used a lower value of lambda than indicated for the anaerobic pond. Was this to be conservative? I was more conservative than you elsewhere, and assumed only 85% cumulative reduction in BOD for the Anaerobic and Facultative ponds. Note that section 6.5.1 suggests that the BOD loading to the first maturation pond should be checked. Note that the equations for total and ammoniacal nitrogen reduction are for Facultative and Maturation ponds. Anaerobic ponds are not mentioned.
Referencing	5%	Generally well-referenced, and in the preferred style.
English		No comments.
Other		Note that there is no single right answer to this exercise.

Essay: How does the altitude of wastewater treatment plant relative to sea level influence certain aerobic biological treatment processes?

(i) Altitude Vs Temperature

Altitude is the level of a place above sea level (in metres). The lower the altitude, the higher the temperature (i.e., the two are inversely proportional) (Afullo A O, 1995). However, temperatures are inversely proportional to the level of dissolved oxygen in surface waters, including wastewaters (Cairncross S and Feachem R, 1983). Similarly, the higher the temperature the higher the reaction rate. The reaction rate changes twofold with a 10^0C change in the temperature. Similarly, the higher the altitude, the lower the temperature. The lower the temperature the lower the reaction rate. The reaction rate changes twofold with a 10^0C change in the temperature. Thus a fall in altitude would increase atmospheric temperatures which would in turn suppress dissolved oxygen levels. However, the increased temperatures would increase growth of aquatic microbes, notably algae, which would photosynthesise and help rejuvenate the oxygen levels of waters (Smith MD (2004). This would increase reaction rates. Therefore, reduced altitudes increase aerobic biological processes; the two are inversely proportional (Crites R and Tchobachnoglous G (1998).

(ii) Altitude Vs Atmospheric pressure (AP)

AP is the force acting perpendicularly per unit area of a surface, such as land. The upper space normally has less material (e.g. gases, dust etc) than the lower parts of the atmosphere. This is partly because the particles would tend to be concentrated in the lower parts due to the gravitational force. It is these particles which exert pressure against a surface they happen to concentrate in / over. On earth, this reflects itself in the form of atmospheric pressure (Smith MD (2004) and Crites R and Tchobachnoglous G (1998).

As the altitude reduces, the atmospheric pressure increases since there are more particles closer to the ground for reasons attributable to the force of gravity (gravitational force).Thus as the pressure increases, so does the concentration of atmospheric oxygen $[O_2]$. Since aerobic treatment processes rely on atmospheric oxygen, this increases their rates. Thus lower altitudes have higher reaction rates for aerobic reactions (Cairncross S and Feachem R, 1983). The reverse is true for higher altitudes, as this is a level at which there is minimum density of atmospheric gases, oxygen included (Crites R and Tchobachnoglous G (1998) and (Smith MD (2004)

Essay: Discuss the important activities that would be performed to help prepare recommendations for onsite effluent treatment options.

Deciding on on-site effluent treatment options require an Audit the inputs and outputs in terms of:

2. Water quantity; determines the effluent flow characteristics, especially volume of effluent, as up to about 80% of incoming water goes out as effluent. Some treatment facilities can handle large volumes of effluent, while others need only small volumes);
3. Land availability assessment: Large tracts of land would most favour waste stabilization ponds (WSPs);
4. Location with respect to the town or residential area: If upwind and there is still enough land for expansion, then WSPs most favoured.
5. Location: If in a hot tropical environment, WSPs most favourable)
6. Nature of outputs: if rich in poisonous substances which would intoxicate the micro flora during treatment, pretreatment would be preferred; if rich in BOD (i.e. very strong effluent, it would dictate the nature of treatment facility)
7. Energy requirement per unit of product (quantity per unit product); some release hot products and effluents, necessitating use of special types of facilities;
8. Quantity of primary Chemical inputs;
9. Nature of primary chemical inputs (quantity, source, toxicity, etc);
10. Nature of additives (e.g. catalysts – biological/ enzymatic or chemical) and their final products in the effluent; reusability option);
11. Nature of products (quantity, toxicity, bio-degradability; alternative uses);
12. Quantity of products (quantity, toxicity, biodegradability, alternative uses);
13. Scale of operation: If scale low / small, then conventional methods preferred if high, waste stabilization ponds (WSPs) would be preferred.
14. Quantity and quality of byproducts;

NB All these depend on the proposed size of the plant. The plant size also depends on the family size / town population, land availability etc. This information would be useful in determining the best method for effluent disposal and treatment. The data can be acquired using the method below. Selection of

appropriate on-site effluent treatment option therefore involves site assessment and evaluation, and can be looked at as two to three separate steps:
Site identification; Preliminary site evaluation and Detailed site assessment.

Preliminary site identification: Involves a study of soils, slope and drainage characteristics. Aspects such as soil depth, permeability, texture, structure etc are assessed. The soil profile is studied, groundwater depth, existing vegetation, surface drainage, and landscapes. The vegetation can be an indication of a type of soil, or some mineral, or some unique water depth and / or characteristics. Pose size, texture and structure can give an indication of pollution prevention potential of the soil, and the need for a surface lining if the place has to be used at all (Crites R and Tchobachnoglous G (1998).

Preliminary site evaluation: According to Crites R and Tchobachnoglous G (1998), this stage involves further tests once a site has been identified. This stage involves determination of current and proposed land use, as well as expected flow, characteristics and use of the effluent. It also involves further detailed assessment of site characteristics. Similar aspects as in stage 1 above are assessed. Hereafter, the relevant regulatory institutions / agencies are contacted for their input / approval. Regulatory factors are used as final criteria. However, further details such as depth to groundwater in the wettest period, and permeability tests are done. Once the site passes regulatory assessment, it undergoes a final rigorous stage of assessment.

Detailed site assessment: Involves detailed study of soil-physical, chemical, biological characteristics, with a focus on the physical aspects. Pit studies are most useful, and official sampling done to determine particle size distribution (texture), soil moisture content, soil depth, and soil structure as it varies with depth etc. Piezometers are used to assess the groundwater fluctuations; percolation tests; allowable hydraulic loading rate (Crites R and Tchobachnoglous G (1998).

According to WEDC (2003), the key questions to ask are: (i) What are we to remove, and (ii) Why these have to be removed.

Basically, what to remove depends on the nature of effluent-industrial... etc. The tanning process will produce strong wastes rich in organic matter, fats, inorganic chemicals, sulphides (and therefore posing a smell hazard) etc.

The key criteria to use to answer the above two questions are: the effluent quality (strength and composition), the availability of funds, available capacity (e.g. technical and professional), site location (which determines whether it has power or nor, and if so, its regularity), land availability, appropriateness of the technology (culturally accepted, financially promising, economically viable, technically feasible), running costs, nuisance characteristics (e.g. smell, dust, noise, fly etc), sludge characteristics, and environmental-health-reuse-discharge standards. The last point refers to the extent to which the technology is able to treat wastes to meet disposal standards, health impacts, and possibilities of reusing the effluent/ or recovering some material from it (WEDC, 2003).

Conclusion: Therefore important activities to be performed will be largely: (i) Study of Soil physical-chemical-biological characteristics (ii) the site hydrogeology (iii) the effluent physical-chemical-biological properties (iv) financial base of the organisation (viii) site location and availability of land.

Explain under what circumstances anaerobic treatment methods may be used in preference to aerobic treatment methods of wastewater treatment

Introduction: Anaerobic treatment methods are known to be cost-effective, as relatively significant reduction in BOD can be realized with minimum cost. It does not require any power for the processes to continue unlike the aerobic processes which require huge amount of power to aerate the system (Cairncross S and Feachem R, 1983). Due to the effectiveness of anaerobic systems, they are used in unique and cruel situations where difficult effluents are involved (Crites R and Tchobachnoglous G (1998). These involve situations such as:

(i) Where BOD is very high such as from food processing and domestic systems which have large quantities of degradable matter. In this situation, it is assumed the effluent in question is very strong (Smith MD (2004). It is extremely expensive to stabilize high BOD effluents, as they demand too much oxygen, often creating heavy oxygen demand and rendering most aerobic microbes ineffective. Furthermore, the natural processes of oxygen dissolution such as turbulence by wind and oxygen dissolution via the air-effluent interphase are often too slow to meet the oxygen demand by the microbes. This often leads to anaerobic system which leads to smelly gases such as H_2S being produced, among others (Crites R and Tchobachnoglous G (1998).

(ii) Where the effluent is relatively rich in toxic chemicals such as heavy metals, solvents, ammonia, antibiotics and pesticides, which are likely to incapacitate aerobic microorganisms; anaerobic systems are able to detoxify such chemicals via some organisms using them as substrates from which they derive their energy, and in the process making them less toxic and more stable (Cairncross S and Feachem R, 1983) and (Crites R and Tchobachnoglous G (1998). Some reduced gases may result. The use of some of these as substrate / food source is useful, as it reduces the sludge volume and toxicity.

(iii) In remote areas where regular power supply and / or use is not feasible, either because there is no power, or because it is expensive. The use of aerobic system in such situations would necessitate the use of a lot of power, which may be unavailable or unreliable. Anaerobic systems thus save the situation due to its low energy demand.

(iv) When resources are limited. In this situation, the costs of effluent treatment are significantly reduced by use of anaerobic systems which are cheaper to run albeit with the disadvantage of noxious smell (Smith MD (2004) and Crites R and Tchobachnoglous G (1998).

(v) Extremely high pH situations. In this case, the normal pH range of 6.5 - 7.5 is long exceeded or underceeded, rendering most microbes ineffective (Cairncross S and Feachem R, 1983). Since the anaerobic microbes are more versatile, they can cope and rectify the situation by using some of the materials to which the extreme pH is attributed to buffers which then stabilize the pH. Thereafter, more organisms can multiply in the more conducive environment, including aerobes.

(vi) Under non-mesophilic conditions, i.e. when temperatures are unfavourable for most microbes. In this case, the high temperatures (common in some industrial effluents) are even associated with suppressed dissolved oxygen in the effluent, a condition under which aerobes will all have died (Afullo A, 1995). The thermophilic anaerobes can cope even in such extreme situations until the temperatures have stabilized when more sensitive microbes start coming in, but in much stabilized effluents. Normally warm effluents interfere with settling process, causing eddies. This can be reduced by having a system with a long retention time, as in the case of anaerobic system.

(vii)In solid-rich effluents, the anaerobic processes will produce less sludge due to the elaborate processes of Hydrolysis, Acidogenesis, Acetogenesis and Methanogenesis. These produce gases from the solid component of the effluent, tremendously reducing the sludge volume. This can reduce the desludging costs (Smith MD (2004) and Crites R and Tchobachnoglous G (1998). It can be therefore very applicable in remote rural situation where desludging capacity can be tricky and scarce.

HYDRAULIC RETENTION TIME AND SOLID RETENTION TIME
Explain why for anaerobic processes, hydraulic retention time (HRT) should be different from solid retention time (SRT). Which is longer, and why:

Introduction: Solid hydraulic time (SHT) is the time (in days) taken by solid components of an effluent in a tank (i.e. the time between entry and exit). Liquid retention time (LRT) is the time taken by any liquid component of the effluent within a tank / reactor, from entry to exit from the reactor. Due to the physical differences between the solid and liquid components of the effluent, they need to be assigned different retention times. In general, and mostly, SRT is longer than HRT (Crites R and Tchobachnoglous G (1998). This is because of the following:

Physical state: Solids may comprise complicated constituents (e.g. polysaccharides, cellulose, proteins, complex starch, fats, grease etc). These solids have to undergo all the four reaction stages within the reactor - including hydrolysis, Acidogenesis, acetogenesis and methanogenesis. Soluble components, now in the liquid form, may have been hydrolysed, thus may need less time as they do not have to undergo all the four stages Smith MD (2004) and (Crites R and Tchobachnoglous G (1998).

Repulsive forces: Since colloidal solids within the effluent are too small, they remain suspended together with the microbes. The insulation in anaerobic reactors caused by thick scum keep the temperatures relatively high. This cause eddies, which further keep most solids in suspension (Crites R and Tchobachnoglous G (1998).

Charges of particles: Most particles in natural systems are negatively charged (Cairncross S and Feachem R, 1983) and (Crites R and Tchobachnoglous G (1998). This includes colloidal suspended solids in effluents, and the microbes. Due to their small size, both microbes and fine solids tend to remain in suspension, repelling one another as they have the same negative charge. There are thus minimal chances of collision and therefore reaction. This keeps the solids in their original inert forms for long. For the solids to be decomposed, they need to take longer in the reactor (high SRT) to increase chances of hydrolysis and possibility of their negative charges becoming weakened in due course. This higher SRT may increase chances of solid breakdown (Smith MD (2004).

Nature of microbes: Some of the most important microbes in the anaerobic system, such as filamentous bacteria, are associated more with breakdown of soluble components of the effluent (Crites R and Tchobachnoglous G (1998). Thus it is easier to break down hydrolysed components, and less easy to decompose solids. The solids thus need more time to increase chances of being broken down.

Reproductive efficiency: Microbes in the solid phase tend to reproduce more slowly. This could be partly due to the repulsion between the microbes and the solid part of the effluent. This significantly reduces the rate of decomposition of solids, necessitating their higher retention time (Crites R and Tchobachnoglous G (1998). Since the solid part of the effluents pose a sludge dimension to effluent management, it is necessary that any means possible should be used to reduce the amount of sludge. This can only be done by increasing the SRT, while trying to optimize reaction conditions within the reactor. Anaerobic reactors thus tend to have longer retention times on the average, with the sole aim of sludge minimization. This targets the solids. Thus any means that can help reduce their volume is applicable and useful, including system designs with higher SRT than HRT.

Conclusion: It is clear that SRT should exceed HRT. To ensure this, sludge or biomass is normally held back or recycled in most reactors, while the liquor is released. Furthermore, most anaerobic reactors are warmed a little to mesophilic range where most microbes operate optimally. This increases the reaction rates significantly. This may help reduce both the SRT and HRT, but the former remains longer.

ANAEROBIC DIGESTION
Essay: Explain how gas produced during anaerobic digestion can obtain carbon in both its oxidized form (CO_2) and reduced form (CH_4).

Generally, according to Crites R and Tchobachnoglous G (1998), the effluent undergoes four stages of chemical breakdown. These include:

Breakdown / digestion Stages
Hydrolysis whereby the solids are broken down into simpler forms via the mediation of water (i.e., water is a reactant). (ii) Acidogenesis: A process whereby the hydrolysed materials e.g. polypeptides and peptides, disaccharides, starch, etc are broken down into large carboxylic acids and alcohols; (iii) Acetogenesis: whereby the complex long chain carboxylic acids are broken down into simple two-carbon organic acid (acetic acid); and (iv) methanogenesis- a process whereby the simple products of acetogenesis are utilized to produce both reduced and oxidized carbon compounds.

These reactions are sequential such that products from one stage are the reactants in the next stage. The reactions thus build onto one another, and no single product is likely to accumulate relative to others. In methanogenesis, the microbes involved are substrate-specific. Thus those using hydrogen or methylamine as their substrate are likely to produce very reduced forms of carbon, such as methane (CH_4), whereas those utilizing oxygen-containing product of acetogenesis such as acetic acid (CH_3COOH) or methanoic acid ($CHCOOH$), are likely to produce oxygen-laden products of carbon (such as CO_2). It is therefore possible for any kind of product to result from anaerobic processes.

SLUDGE MANAGEMENT IN ANAEROBIC DIGESTERS
Explain why for sludge flowing out of a heated anaerobic digester, it may be difficult to separate solids from liquids using the conventional physical methods.

Introduction: The conventional physical methods of separating solids from liquid parts of sludge include (i) centrifugation and

(ii) filtration of various forms e.g. vacuum and filter pressing. Also, generally, the anaerobic reactors are lightly heated to the optimal mesophilic reaction range (about 35^0C), to improve the reaction efficiency (Cairncross S and Feachem R, 1983); (Smith MD (2004) and Crites R and Tchobachnoglous G (1998).

Particle Size: Size of most common sludge components are solids of diameter 1-5 µm (Cairncross S and Feachem R, 1983). These are the most difficult to dewatering, at whatever temperature. This is partly due to the high cohesive, plus adhesive forces, in which the former hold the water particles together, and the latter force holds the water and particulate particles together. Thus the close proximity between the water particles and the colloidal matter / fine solids in the sludge makes it extremely difficult to dewater. Some of the water is even part of the structure of the settled sludge, giving it the stable and crumb structure it usually takes. This stability makes it difficult for any water particle to be released. Further, the capillary water held in the sludge is too strongly held that the spaces in between does not allow them to be released easily. They may only be forced out using stronger forces than the conventional dewatering techniques (Mara D and Pearson, H (1988) and Smith MD (2004).

Speed / rate of formation Vs stability: Sludge resulting from heated anaerobic processes is formed too slowly, often at the process-efficient mesophilic range. This slow process produces the sludge very slowly, and the product is very stable. It is often difficult to change the physical or chemical form of any stable substance, including removing any component of it. This applies to attempts to dewater the sludge (Smith MD (2004). Generally, anaerobic processes increase moisture content of effluents by a factor of 2; improve dewaterability and reduces settleability. This efficiency at an earlier stage makes it difficult to dewater the more stable product (Crites R and Tchobachnoglous G (1998) and (Mara D and Pearson, H (1988).

Additives and chemical composition: In some instances, lime is added to effluent in reactors to modify its pH to the appropriate and efficient range. The resulting sludge is also most likely to be rich in calcium and other polycations. These increase the osmotic pressure of the sludge, thereby making it almost impossible to separate the water held therein. Calcium has very strong cementing properties (Smith MD (2004) and Crites R and

Tchobachnoglous G (1998)) and normally able to stabilize compounds and materials such as soil and sludge It is naturally difficult to distabilise a stable product. In this case the sludge resulting from a limes effluent is equally difficult to extract anything from, leave alone manipulating in any way (Crites R and Tchobachnoglous G (1998)).

Conclusion: Therefore for sludge flowing out of a heated anaerobic digester, it is difficult to separate solids from liquids using the conventional physical methods.

Essay:Scenario as above; calculation of [solids] in sludge after dewatering is required.

The method used is a mass balance idea adopted from ((Crites R and Tchobachnoglous G (1998), WEDC (2003) and WEDC 2003b)
26 litres of water to be removed from the sludge.
Initial weight of sludge = 62 kg;   Final weight of sludge = 62 – 26 kg = 36 kg;   Sludge moisture content = 97.5%; solid content = 2.5%; Mass of solid in sludge prior to dewatering = 2.5% of 62kg = 1.55 kg
Since this same mass of solid remains after dewatering, the final % of solid in the sludge = $1.55 / 36$ x 100 = 4.31%; Therefore the moisture content of sludge after dewatering is 100 – 4.31% = 95.69%

Essay: (a) Calculate the BOD leaving the treatment plant in each case (WEDC (2003) and WEDC (2003b)

Complete mix model	Plug flow model
Certain volume of effluent emerges from inlet pipe in unit time and is immediately mixed and distributed evenly throughout the reaction vessel. WEDC (2003) and WEDC (2003b)	Is a batch process whereby a certain volume of effluent emerges from the inlet pipe in unit time and flows through a reactor vessel without mixing with other portions of wastewater flow (WEDC (2003) and WEDC (2003b)
Applicable in waste stabilization ponds (WSPs) and aerated lagoons.	Applicable in a sewage standing for a long period of time
Equation: $Le/Li = 1/{1+Kit}$	Equation: $Le/Li = e^{-kit}$
Le = BOD leaving the waste treatment plant in mg/l	Le = BOD leaving the waste treatment plant in mg/l
Li = BOD of incoming effluent in mg/l = 185 mg/l	Li = BOD of incoming effluent in mg/l = 185 mg/l
K_T =0.23 day^{-1} at 20^{0}C	K_T =0.23day^{-1} at 20^{0}C
T = retention time in hours = 13 hours = $13/24$ days	T = retention time in hours = 13 hours = $13/24$ days
To calculate BOD leaving the waste treatment plant (Le), when The BOD of incoming effluent (Li) is 185 mg/l,	To calculate BOD leaving the waste treatment plant (Le), when The BOD of incoming effluent (Li) is 185 mg/l,
$Le/Li = 1/{1+Kit}$ (Crites R and Tchobachnoglous G (1998).	$Le/Li = e^{-kit}$ (Crites R and Tchobachnoglous G (1998).
	Le = Li e^{-kit} = Li /e^{kit} = $185 / e^{.23 \times 14/24}$
Le = $Li/{1+Kit}$	= 185 / e$^{.12458}$ = $185 / 1.132$
Le = $185/{1+0.23t}$ = $185/{1+(0.23 \times 13/24)}$	= 163.3 mg/l
= $185 / {1+0.125}$ = $185 / 1.125$	
WEDC (2003) and WEDC (2003b)	(WEDC (2003) and WEDC (2003b)
= 164.4 mg/l	
The effluent BOD at 20^{0}C under complete mix model is 164.4 mg/l	The effluent BOD at 20^{0}C under plug flow model is 163.3 mg/l

NB: Values of Ki can be determined at $20^{\circ}C$; can be corrected to temperatures within the range of $5 - 25^{\circ}C$ using the equation: $K_T = K_{20}Q^{(T-20)}$ where Q is Arrhenius constant with values from $1.01 - 1.09$; ((Mara D and Pearson, H (1988), (Smith MD (2004), WEDC (2003) And WEDC (2003b).

Essay 6(b): what effect would an increase in temperature have on the value of K_1?

Taking an example of the complete mix model, the equation below applies. The temperature is not reflected directly in the equation, but can be inferred since it affects the rate of reactions almost in a uniform pattern. Thus its increase would decrease the BOD (thus reduce the Le/Li). This line of reasoning is furthered below:

Using the equation $^{Le}/_{Li} = {}^1/_{1+Kit}$ ((Crites R and Tchobachnoglous G (1998); WEDC (2003) and WEDC (2003b)).

Generally, $^{Le}/_{Li}$ is inversely proportional to temperature WEDC (2003) and WEDC (2003b). Thus as the temperature increases, the extent of degradation of matter in effluent (BOD decline) increases. A fall in the left hand fraction (LHS) means there must be a fall in the right hand fraction commensurate with the decline in the LHS. With the numerator (1) being a constant, and for the fraction $^1/_{1+Kit}$ to decline, there must be an increase in the figure at the denominator.

Thus 1+Kit must increase proportionately. Since 1 is a constant here, the only thing that can increase 1+Kit is Kit itself. However, the value of t is constant. It is only logical that Ki must increase in the circumstances (Mara D and Pearson, H (1988). Therefore as the temperature increases, there must be an increase in Ki. The two are therefore directly proportional.(Crites R and Tchobachnoglous G (1998), WEDC (2003) and WEDC (2003b)

This reasoning applies even in plug flow situation whose equation is given by: $^{Le}/_{Li} = e^{-kit}$ Smith MD (2004). As the LHS declines (due to temperature effects as explained earlier), $1/e^{kit}$ must decline. Thus e^{kit} must increase. The only logical way of increasing the figure of e^{kit} is by increasing its power. But since the power comprises a variable and a constant, it is only logical that the variable increases commensurately. Thus for LHS to decline, Ki must increase. This depicts, once more, a direct proportionality between the temperature and Ki in any kind of effluent flow (Mara D and Pearson, H (1988).

REFERENCES

a.   Afullo A O (1995). Pollution of Lake Victoria by inorganic fertilizers used in West Kano Rice irrigation Scheme, Kisumu, Kenya. An unpublished MPhil Thesis, Moi University, Kenya.

b)   Cairncross S and Feachem R (1983). Environmental Health Engineering in the Tropics: An introductory Text 2nd edn. John Wiley and Sons, Chichestershire.

c)   Crites R and Tchobachnoglous G (1998). Small and decentralized wastewater management systems. McGraw Hill International editions. Singapore

d)   Mara D and Pearson, H (1988). Design manual for Waste water stabilization ponds in Mediterranean countries. European Investment bank and Mediterranean Environmental Technical assistance Program, in collaboration of Lagoon Technology International, Leeds, England

e)   Smith MD (2004) Wastewater Treatment: July 2004 topic task 2. WEDC, Loughborough University, UK.

f) WEDC (2003) Wastewater treatment: A post-graduate module. Loughborough, UK.

(g) WEDC (2003b) Wastewater treatment. Additional resources. Loughborough, UK.

WASTEWATER TREATMENT TOPIC TASK 3

Alternative / suggested / Outline Responses.

1. Explain how the altitude of a wastewater treatment plant relative to sea level may influence certain aerobic biological treatment processes.

Altitude affects PRESSURE, which influences solubility of oxygen and the amount of oxygen per unit volume. At higher altitudes, less oxygen can be dissolved in wastewater, so aeration equipment is less efficient, and the oxygen available in wastewater is limited. As oxygen is used by aerobic organisms, it will need to be replaced, but oxygen transfer becomes less efficient with increasing altitude. The effect of this will be to slow down aerobic treatment processes. Nitrification will also be adversely affected.

Altitude may also affect Temperature (temperatures fall as altitude increases). The effect of low temperatures is also to slow down biological treatment processes. Differences between day-time and night-time air temperatures may be considerable. As temperatures drop, the saturation concentration of oxygen in water increases, but the diffusion rate decreases.

The overall effect of temperature and pressure on transfer rates, metabolic activity and saturation concentrations is complicated, but overall aerobic treatment processes slow down at high altitude.

Reduced solubility of CO_2 may affect pH, influencing microbial activity.

Algal production of oxygen in waste stabilisation ponds may increase slightly if the atmosphere is clear at high altitude, because of solar radiation. This will depend on water temperature and availability of nutrients.

There may also be increased rainfall at high altitude, possibly diluting the wastewater to be treated.

Low temperatures will affect physical treatment processes associated with the viscosity and density of the water. Sedimentation rates will reduce at low temperature, but sedimentation is not an aerobic biological treatment process.

(Discussion of these points could be applied to specific treatment processes.)

2. Managers of a small tannery (leather-preparing) industry wish to provide some on-site treatment for the wastewaters produced. List some of the important activities that you would perform to help you prepare recommendations for treatment options.

    Assume that the tannery processes use a variety of organic and inorganic chemicals, and that the wastewater streams will contain a variety of corrosive and hazardous chemicals. You are not expected to know what chemicals may be used in the tanning process, or to know about tanning activities.

3. Wastes may be discharged into a sewer or into a stream. Treatment or pre-treatment may be necessary for either discharge option. The very fact that the industry is considering treatment suggests that they are not considering illegal disposal of wastes. Some people began by considering: "What are we trying to remove? And why?".

    Important considerations for sustainable wastewater management are:
    - Economic justification for the choice of treatment
    - Financial viability (affordability)
    - Technical feasibility

    Specific points to be considered.
    Consider possible discharge points and options; and the costs for different discharge options.
    Consider effluent standards required at discharge.
    Consider possible environmental impacts of discharges.
    Consider land available for treatment.
    Consider business plans for the industry in the future.
    Consider the economic situation of the company.
    Identify all chemicals used.
    Identify all processes producing wastewaters, and identify inputs and outputs. Also attempt to assess whether

quantities of any materials (especially water) could be reduced.

Identify, sample, analyse and characterise all wastewater streams (quantities and contents), possibly including measurement of the temperature of individual streams.

Measure flows into and out of the works.

Account for any discrepancies between flows in and out.

For each stream determine whether it is best treated separately, or blended with other streams before treatment. Determine whether balancing tanks could be used to reduce variation in flows.

For each stream determine appropriate forms of treatment (physical, chemical, biological – either aerobic or anaerobic).

Investigate whether similar industries in the region already use appropriate treatment processes.

Decide whether batch-treatment or continuous treatment is more suitable for individual streams.

Establish whether there are any alternatives to on-site treatment.

Establish whether any wastewater streams can be re-used or recycled.

Establish whether it is possible or desirable to recover any materials.

Establish how waste characteristics vary, during the day or seasonally. (I was once involved in a project where wastewater treatment became difficult because an industry closed down for about three weeks at the coldest time of the year.)

A desk study, and visits to other tanneries, would provide useful background information about treatment options, treatment costs, problems encountered, suitable construction materials, etc.

3. a. Explain under what circumstances anaerobic treatment methods may be used in preference to aerobic treatment methods for treatment of wastewaters.

b. For anaerobic treatment processes, explain why it is desirable for the Hydraulic Retention Time (HRT) to be different from the Solids Retention Time (SRT). Your answer should state, with reasons, which of the two (HRT or SRT) should be longer.

c. Explain how the gas produced during anaerobic digestion can contain carbon in both its oxidised form (carbon dioxide, CO_2) and in its reduced state (methane, CH_4).

d. For sludge flowing out of a heated anaerobic digester, explain why it may be difficult to separate solids from liquids using conventional physical methods.

3. Anaerobic treatment may be used for:
   - strong wastes (difficult to treat)
   - problem wastes, possibly deficient in some nutrients (difficult to treat)
   - where oxygen cannot be supplied sufficiently quickly
   - to reduce volumes of sludge
   - to produce useful materials (such as biogas)
   - where power is not available
   - where treatment plants may not receive flows for long periods (days, weeks or months)
   - where there is limited land available
   - where waste temperatures are high

Common applications are for treating wastes in the food industry, and for treating wastewater sludges. Anaerobic treatment units generally need to be larger than aerobic units. Some domestic wastes may receive anaerobic treatment, for example in septic tanks or in anaerobic waste stabilisation ponds. Anaerobic treatment is only partial, producing products which can be broken down into simpler chemicals. Anaerobic treatment often needs additional treatment.

SRT should be longer than HRT.
Some units have HRT and SRT the same, but it is desirable to keep SRT greater than HRT. The HRT needs to be kept to a minimum, to keep the reactor volume small, but SRT needs to be greater because anaerobic bacteria do not multiply quickly. Biomass is lost if discharged, and cannot be replaced quickly. A high SRT also provides extra protection during periods of high loading, and provides some insurance against any inhibition of anaerobic activity.

c. Carbon dioxide, although an oxidised form of carbon, can exist because it has a low solubility and rapidly floats to the surface as a gas. It will appear as a gas unless it combines with other chemicals to form (for example) bicarbonates. Every oxidation reaction is balanced by a reduction reaction, and vice versa, and the reduction of carbon to methane is accompanied by oxidation, so carbon dioxide is also produced. The oxygen comes from water (H_2O).

d. Sedimentation may be difficult because the solids are light and may be attached to gas bubbles. Much of the water may be bound, so that it is not free-draining. Sedimentation may also be difficult because convection currents in the warm fluid help to keep solids in suspension. At warm temperature the liquid will have low viscosity, affecting convection and settlement. Dissolved gas flotation is difficult because air cannot be used (it contains oxygen) and methane is inflammable, and may explode if compressed.

5. A total of 26 litres of water is removed from a sample of wastewater sludge having an initial weight of 62 kg, and an initial water content of 97.5%. What will be the new water content of the sludge after removal of the 26 litres of water?
   Note: Water content = (Weight of water) ÷ (Weight of water + Weight of solids)

6. Let W_w and W_s) represent weights of water and solids respectively.
   Initially $(W_w + W_s)$ = 62 kg
   Initial water content = $(W_w) ÷ (W_w + W_s)$ = 0.975
   $(W_w) ÷ (62)$ = 0.975
   (W_w) = 60.45 kg
   (W_s) = 1.55 kg
New water content = $(60.45 - 26.0) ÷ (62.0 - 26.0)$ = $(34.45) ÷ (36)$ = 0.957
   New water content = (95.7%)

7. Figure Q5 shows how concentrations of dissolved oxygen increased when air was bubbled through a sample of distilled water at a temperature of 20°C. Initially the water sample contained no dissolved oxygen. An additional copy of Figure Q5 is provided.
   On the additional copy of Q5, sketch similar graphs showing how dissolved oxygen concentrations are likely to increase in each of the following situations:
   (i) A sample of distilled water at a temperature of 25°C.
   (ii) A sample of domestic wastewater at a temperature of 20°C.
Show clearly which sketch graph applies to which situation.
Assume that the sample sizes and equipment remain unchanged, with the only changes being to either the temperature or composition of the water. Also assume that the samples initially contain no dissolved oxygen.
   NOTE: The saturation concentration of atmospheric oxygen in pure water (C_s) at a temperature (T) can be estimated using the equation:

$$(C_s) = \left\{ \frac{468}{31.6 + T} \right\} \quad (\pm 0.04) \qquad \text{where } (C_s) \text{ is in mg/l and T is in °C.}$$

Figure Q5.

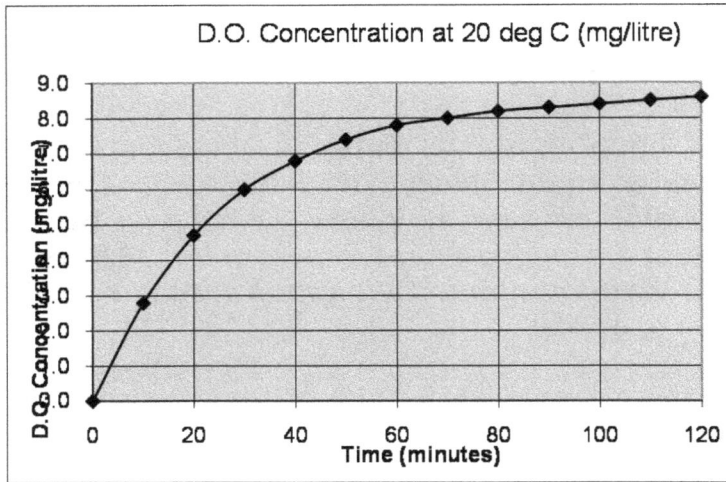

D.O. Concentration at 20 deg C (mg/litre)

5. Note that this question is about solubility of oxygen, not about the dissolved oxygen sag curve.

For the distilled water at 20°C: $\quad$ Cs = 9.07 mg/l

For the distilled water at 25°C: $\quad$ C_s = 8.27 mg/l

For distilled water at 25°C the graph will have a similar shape to that shown in Figure Q5, but the initial slope will be steeper (oxygen transfer rates increase as temperature rises), and the curve will then flatten out below the original line. The new line will therefore be above the original line initially, but the lines will then cross and the new line will tend towards a value of 8.27 mg/l.

For wastewater at 20°C: $\quad$ C_s = 0.85 × 9.07 mg/l = 7.7 mg/l

Wastewater has a lower saturation concentration, and a slower oxygen transfer rate, so the curve will remain below the original line shown in Figure Q5, approaching a value of 7.7 mg/l.

6a. Two ways of modelling BOD removal kinetics are the "complete-mix method" and the "plug-flow method".

For an aerobic biological treatment plant having a retention time of 13 hours, use each equation to calculate the BOD leaving the treatment plant. Assume a BOD of 185 mg/l entering the plant, and assume that k_1 = 0.23 days^{-1}.

What effect would an increase in temperature have on the value of k_1?

a.

$$\left\{\frac{L_e}{L_i}\right\} = e^{-k_1 t^*} \quad \textbf{Plug flow}$$

$$\left\{\frac{L_e}{L_i}\right\} = \frac{1}{1+k_1 t^*} \quad \textbf{Complete mix}$$

$L_e/L_i = e^{-k1t}$ $\qquad\qquad$ $L_e/185 = e^{-0.23 \times (13/24)}$

$L_e = 185 \times e^{-0.23 \times (13/24)} = 163.3$ mg/l

$L_e/L_i = 1/(1+k_1 t)$ $\qquad\qquad$ $L_e = 185/(1 + (13/24) \times 0.23)$

$L_e = 164.5$ mg/l

An increase in temperature will increase k_1, reducing the time for treatment.

SECTION 4: WATER FOR LOW INCOME COMMUNITIES

TOPIC TASK 1
POPULATION PROJECTIONS

Essay 1.1   : If you only had the information provided above, what would you estimate the present population of the village to be?

According to Skinner (2004), if the current population is N_0, the future population (N_t) after t years is given by the equation:
$N_t = N_0 (1+p)^t$ Where p is the annual population growth rate; t = years elapsed; N_0 = the current population; and N_t is the population after t years. In our case, N_0 =516;   p= 2.5%; t = 2005 – 1995 = 10 years; 1+p = 1.025; N_{10} = 516 x $(1.025)^{10}$ = 516 x 1.291 = 667. Therefore the current population (i.e., in 2005) is 667.

Assumptions: Census figures are normally taken as the most reliable, and should be used when available (WEDC, 2004). Since immigration and HIV / AIDS balance out, the figure of 516 is assumed for the area, and the population growth rate is assumed to also balance out. It is assumed to offer some margin of safety. The figure of 600 (100 households@ 6 members) could be an overestimate from the village elder.

What do you estimate the population to be in 2020? Clearly show your calculations and comment on factors that would affect the reliability of each factor? (10%) (Ref: Skinner, 2004)
$N_t = N_0 (1+p)^t$ Nt = N_{25} = 516 $(1.025)^{25}$ = 516 x 1.897 = 998. Therefore the estimated population of the village in 2020 is 998.

Assumptions: Census figures are normally taken as the most reliable, and should be used when available (WEDC, 2004). Since immigration and HIV / AIDS balance out, the figure of 516 is assumed for the area, and the population growth rate is assumed to also balance out. It is assumed to offer some margin of safety. The figure of 600 (100 households@ 6 members) could be an overestimate from the village elder.

Question 1.2: In question 1.2, assume that just one method of water supply is to be chosen for the whole village, and that there are only three feasible options:

A: A piped supply of spring water from a storage tank. The water will be delivered to public standposts (tapstands) distributed through the village. The system will provide water for all domestic needs.

B: A number of handpumps on boreholes distributed through the village. The system will provide water for all domestic water needs.

C: household rainwater catchment systems (using house roofs and a storage tanks at each dwelling). The system will provide water for only drinking and cooking; other water needs will be supplied by existing sources.

TASKS/ ESSAYS:

i) Does the level of accuracy required for estimating the village population needs to be different to effectively design each of these methods of supply? Clearly explain the reasons for your answer. Restrict your answer to ¾ of a side of A4-sized paper. (5%)

Yes, the level of accuracy required for estimating the population needs to be different for each of these water supply methods. Each method depends on a number of factors, but all depend on: stability of the population and development. (i) Where stability exists, design period of more than 10 yeas is possible. (ii) Materials and equipment: The working life of the materials used for each system dictate the design life of the system. (iii) Layout: In places where future developments are predictable, it is possible to use a longer design period, which then does not need a very high level of accuracy in population estimates.

The above is because a larger factor of safety must be inbuilt into the system to accommodate any likely changes over the long design life. The reverse is true for shorter design periods. In unpredictable systems, looped water distribution is preferred to branched systems due to varying levels of adaptability (Skinner, 2004). However, the latter can still be converted to a looped system if the design life is long, and changes are warranted before the end of the design period. (iv) Construction, operation and maintenance: These are required, and where funds are unavailable, shorter design life is better, and thus estimates of population need to be shorter term. These are explained by option below:

Option C is not very feasible, and the least likely of the three options, given the variation in socio-economic characteristics of individuals and households. Households are likely to have different roofing material, thus making it difficult to collect clean roof water in sufficient quantities. Option B is moderately feasible, rating second of the three options, given the need for pumps to be bought, maintained and repaired. There are likely to be times when the pumps are non-operational due to temporary failures, and requiring repair or maintenance. Since pump lives are normally 5-10 years, the design life of less than 10 years looks reasonable (Skinner, 2004). The use of 10 or 25 years are unreasonable for pump systems. For this option, the population estimate needs to be very exact, as it is meant to guide design period more accurately.

Option A looks most appropriate for the area. This is partly because the population growth is stable and can be predicted. Pipes can last more than 50 years, and this fits well with a village with such stable demographic characteristics (Skinner, 2004). Thus design population can even be of 50 years or more. For the options with longer design periods, the population estimate needs not be too exact, as a lot of factors can affect the population long before the design period elapses.

ii) Inaccuracies in predicting future populations and their water demands can sometimes be catered for by applying a factor of safety to the number estimated. Another strategy is to plan for staged construction, so the design can be upgraded at a later date to cope with actual changes in population or water demand. Briefly explain how factors of safety, or staged construction, could be included with each of the supply systems (a, b and c) to make allowance for inaccuracies in predicting the population (3/4 of page).

In unpredictable service areas, looped water distribution is preferred to branched systems due to varying levels of adaptability (Skinner, 2004; Cairncross, s and Feachem R (1993). However, the latter can still be converted to a looped system if the design life is long, and changes are warranted before the end of the design period. Construction, operation and maintenance are required, and where funds are unavailable, shorter design life is better, and thus estimates of population need to be shorter term. Shorter term plans need less safety factor than longer design periods.

Option C is not very feasible, and the least likely of the three options, given the variation in socio-economic characteristics of individuals and households of the village. The roof catchments need the impervious roofing material, gutters, a down pipe and a tank (Skinner, 2004). The initial stage could involve investing in the roofing material and the gutters, followed by a small tank (perhaps the normal household water storing materials such as the 20-litre buckets, 100 litre and 300 litre movable tanks. These can be upgraded to above ground and / or underground tanks later. An underground tank can be dug, and gradually built using bricks and cement, and lead the roof water into the completed lower section of the tank which is constructed upwards gradually until it attains its final height. Meanwhile the completed lower section of the tank continues to be used, while the upper part is under construction. There is no need for any addition of population above the calculated, as the roof catchment are entirely dependent on existing households, and each serves a household. This incremental approach can work in piped system as well. However, it takes time to build the water supply to the demand level (Cairncross, s and Feachem R (1993).

Option B (handpumps in boreholes) is moderately feasible, rating second of the three options, given the need for pumps to be bought, maintained and repaired. There are likely to be times when the pumps are non-operational due to temporary failures, and requiring repair or maintenance (Cairncross, s and Feachem R (1993); and WEDC, 2003). Since pump lives are normally 5-10 years, the design life of less than 10 years looks reasonable (Skinner, 2004). The use of 10 or 25 years are unreasonable for pump systems, thus an extra population of 10% above the projection can be added to cater for this unpredictability. An initial plan of hand pump can be adopted, and later upgraded to a powered pump.

Option A (piped supply of spring water) looks most appropriate for the area. This is partly because the population growth is stable and can be predicted. Pipes can last more than 50 years (Skinner, 2004) and this fits well with a village with such stable demographic characteristics. An extra population of about 10% can be added to cater for any unpredictable demographic characteristics. For piped supplies, storage can be provided as a safety factor. This means that if a pump breaks down, or there is an unexpected peak usage, an extra day's supply equivalent can still be available while the pump is under quick repair (WEDC, 2003; Skinner, 2004; and Cairncross, s and Feachem R (1993). The stored water may also serve during emergencies such as for fire fighting. Incremental provision of storage can make the system more affordable, just like in roof catchment option.

Looped water distribution can first be adapted due to varying levels of adaptability (Cairncross, s and Feachem R (1993). However, the branched system can still be converted to a looped system if the design life is long, and changes are warranted before the end of the design period (Cairncross, s and Feachem R (1993). Construction, operation and maintenance are required, and where funds are unavailable, shorter design life is better, and thus estimates of population need to be shorter term. Shorter term plans need less safety factor than longer design periods (Skinner, 2004)

Essay 1.3: How do you think a more accurate estimate of the present and future population could be obtained in a cost-effective way? List, and briefly describe the different activities that could take place to find these more accurate figures. Indicate any major advantages and disadvantages associated with each method. State which of the activities you would favour and why. Restrict your answer to ¾ of a side of A4-sized paper.

The most effective ways are many, and none may be perfect. It requires use of a combination of methods, some being used for triangulation (Skinner 2004; WEDC, 2003)). Where the past census data exists, it is the best starting point. This can be used, alongside the annual population growth rate for the area, or for a neighbouring area with similar socio-economic characteristics, to estimate the present and future population by extrapolation (Skinner, 2004). Extrapolation can be done by graphical methods or by calculation. This gives the quickest result, is cheap, and reasonably accurate. However, this may not take into account population dynamics such as migrations, as well as epidemics which may have occurred over that period after the last census. The result can therefore be confirmed by doing spot-checks in sections of the area, checking of existing records for any significant immigrations, births and deaths over the period after the census, and triangulation by using key informer interview, which could involve the local population office, health office / records, policy makers and local residents.

In emergency situations, if there is a means, aerial photos can be used to get the number of dwellings / households, and multiplied by the estimated household size to get the estimate total population (Skinner, 2004). However, this may be misleading, as some houses may be for different uses e.g. kitchen and livestock, and no human being may be living there. The result from aerial photo therefore still has to be confirmed by doing a random census in a few households to ascertain the household population. It is expensive, but may give quick and reasonably accurate results. Lastly, a full house and population census can be done. This though is time consuming, expensive and difficult to organize (Skinner, B, 2004). But it gives the most accurate and reliable results.

In general, there is a variety of population estimation. But their disadvantages range from being expensive and slow, to being cheap, unreliable, and difficult to organize (Skinner, 2004). Whatever it is, there is need for at least two methods to be used to ascertain the population estimates for validation purposes (WEDC, 2003). It may be useful, unless under emergency situations, to invest in a reliable census however expensive because its long term use for planning, among others, eventually balance out the immediate short term financial demands.

ESTIMATES OF WATER DEMAND

Essay 2.1: Calculate the total desirable volumes of water estimated for each well to supply each day at the end of the design life (8 years).

The table below gives the water requirements per relevant category (*adapted from Skinner, B, 2004 pp 35; Jordan, pp 29; pp 64 of (Nozaic, 2002) as well as the total water demand for the village in eight years' time.

Table 1: Water demand in the village

Category	Requirement / day lcd*	Number	Total daily water demand (litres)
Man / residential	45	680	30,600
Goats	20	20	400
Cattle	35	100	3,500
Chicken	20l/100	500	100
School	140	400	56,000
TOTAL			90,600

NB: Assume a boarding school with 320 pupils and 80 staff-totaling 400. Cattle are only found in the well 4 catchment, while primary school is only found in well 1 catchment. Thus from the total village daily water demand of 90,600 litres, the unique demands for specific well catchment areas (as shown in the map overleaf), the figure for common demands can be calculated. Thus 90,600-56,000 -3,500 = 31,100 litres.

This daily water demand covers common utilities e.g. human, goats and chicken. It is therefore divided proportionately according to the number of households as counted in the map given. The total numbers of household units counted from the village map are 126. These are distributed as follows: 29,29,23,20 and 25 within catchments of wells 1,2,3,4 and 5 respectively. Proportionate distribution of the water resources gives a little over 247 litres per household per day. This is rounded up to 250 litres per household per day. When the figure of 250 is multiplied by each wells' households, it gives 7250, 7250, 5750, 5000 and 6250 litres for wells 1,2,3,4 and 5 respectively. The total demands from each well (and well catchment) are therefore as shown in the table below.

Table 2: Estimate of total daily water demand in the village

Well	Number of residential units	Other separate / special water demands	Special water demands	Common water demands*	Total daily water demand per well
W1	29	Primary school	56,000	7,250	63,250
W2	29	-	-	7,250	7,250
W3	23	-	-	5,750	5,750
W4	20	Corral / cattle	3,500	5,000	8,500
W5	25	-	-	6,250	6,250
Total	126		59,500	31,500	91,000

NB: Common water demands cover human, poultry and goats; I residential unit utilizes 250 litres of water per day;
The discrepancy between the water demands for the village seen in the two tables (90,600 and 91,000 litres) is because of the rounding off of 247... to 250, used to distribute the water demand from each well. The figure of 91,000 is adopted to offer an extra safety margin in water supply to the village.

Essay 2.2: The water requirements calculated above (Q 2.1) need to not be different if the wells were equipped with a high yielding hand pump delivering 25l/min (36,000 l/day) instead of 20l/min (28,800 l/day).

This is because according to Skinner, B (2004), the maximum rate at which water can be extracted from a borehole using a hand pump is relatively low, such that yields are of up to 0.2-0.3 litres / second (12-18 litres / minute OR 17,280 l/day – 25,920 l/day). Use of higher powered handpumps is likely to lead to larger drawdown, leading possibly to subsidence or even salt instrusion. However, in this case, the hand pump of 20l/min capacity is already functioning beyond that of a low-capacity pump of 18 l/min. Therefore, there may be no need to adjust the water demands to the capacity of the higher-capacity pump. If a hand dug well is used, then even lower average yields than 0.2-0.3 l/sec for boreholes can often still meet the community water demands. This is partly because unlike small diameter boreholes, hand dug wells are able to store appreciable quantities of water; the aquifer can yield water at a low rate to storage throughout the day and night (Skinner, B, 2004, page 4.8).

Essay 3: Question 3.1:
Gravity overflow springs occur when water flows in an unconfined aquifer which outcrops because of a change in ground slope, or because an impervious layer which cuts through the aquifer reaches the ground surface (Skinner, 2004, pg 5.7; and Tayong, 2002). This kind of spring has variable yield, sometimes being seasonal. If springs A and b are that type, with groundwater having intercepted by a nearly horizontal , i-m thick layer of fairly impermeable clay that lies under the sandy soil and sand stratum on the hillside above the village, the water is unlikely to be of good quality. This is because, by virtue of being an unconfined aquifer, there are chances that the soil based pollutants have not been purified (filtered) enough to render the water clean. It is therefore likely to be infested with pathogens, nutrients and chemicals from septic tanks, pit latrines, sewers etc.

Due to a possible short distance in underground, the suspended solids may not even have fully settled, especially in rainy season spring, rendering the spring water turbid. This means that such water is likely to have one or a combination of biological, chemical and physical impurities. It should be treated like well or borehole water prior to use.

Essay Question 3.2: Springs A and B are to be used to supply water to some villages. Besides knowing the daily total demand for water from the springs supply, a designer may need to try to predict the pattern of demand that the system needs to supply over various periods during the day. This is because there are a chance the water from springs A and / or B may be turbid, and if sedimentation overnight can be planned for, the better so that the turbidity subsides while the water stays in the sedimentation chamber. Also, the peak rate of demand is useful to know so that it can be balanced with the rate of flow of the spring (Skinner, B, 2004, pg   and Shaw (1999). Then a decision on whether a storage chamber is necessary or not can be reached. For if then peak rate of demand exceeds the rate of flow of the spring, then the design includes a storage chamber, and the reverse is true (Tayong, (2002).

USE OF PUMPS IN WATER DRAWALS
There are a number of disadvantages of using pumped treated river water compared to the wells scheme with buckets and windlasses. These are discussed in the paragraphs below.

First, the pump may break down, and leaves the community with completely no water. The breakdown may be as a result of non-maintenance, or due to lack of a spare-part. These could be either unavailable or expensive to purchase. This will have an extra demand from the community, and unless they are organized in committees which facilitate the repairs, the community may resort back to old untreated water which is likely to cause more disaster-possibly a water-borne epidemic. The facility may be vandalized if the community does not own the idea, planning, design, implementation, operation, and monitoring. This may lead to the same result of lack of regular and reliable supply, with a possibility of an epidemic.

People are likely to develop too much faith and overconfidence in the treated water, such that even when it is not treated- by commission or omission- then it is likely a crisis may arise like a water borne calamity. This may lead to total lack of confidence in such schemes, leading to a bleak future as the community is unlikely to cooperate if faced with a related scheme. Chlorine may have taste which is not acceptable to some community members. This may lead them to even sabotage the scheme (Skinner, 2004; Shaw, 1999). Improper treatment of the water may lead to chlorine reacting with some chemicals in the water, leading to poisonous products (Skinner, 2004; Shaw, 1999). These may pose health risk to the community. Example is chloramines, if the water has any traces of ammonia, nitrates or nitrites. These are carcinogenic (Afullo, 1995).

There is need for chemicals to be supplied regularly; their availability (enough, in time, and at affordable rates) is useful for the future of the project. Once these are used, there is need for continuous monitoring of the program, especially water quality to be sure it is worthwhile. The system can only work if the software components are addressed appropriately, such as the need for health education so that villagers take care after drawing the water- and protect it from contamination. This aspect is useful, as it informs the villagers of their very vital role if the system is to succeed. Carrying the water to a tank at contour 148 means that water is to be raised by a head of 23 m (148-125m). This may be quite a hard task, and a lot of fuel may be required. Lack of diesel can therefore completely ground the scheme. Also it will require a very strong pump , either a direct action or deep well pumps which can lift up to 25 m or 80 m respectively (Shaw, 1999 pg 34).

Distribution of the tapstands may be a big issue, which may cause disagreements, and some people continuing to use the raw river water as it may still be closer to them (Shaw, 1999; WEDC, 2003. The water may contain chemicals which react and wear out the pump, or the pipes, leading to its early failure. There is therefore a need to monitor the general river water quality to ascertain there is no significant impact on the engine or the pipes. There may be need to even protect the water upstream and uphill by afforestation programs. The success of the pump relies on the success of the filtration system; if the latter fails, the pump is likely to be clogged, leading to its failure, and that of the entire system. This over-reliance of the system on the efficiency of one step / process is too delicate to rely on for a scheme hoping to completely support livelihoods. This is significant for this design because during wet season, the river water is highly turbid. The system is expensive, requiring a lot of capital for pipes (and tapstands), tank, diesel and pump, among others. It is unlikely to pick up unless the community owns it; it is likely to be very expensive.

REFERENCES:
1. Afullo A (1995). Pollution of Lake Victoria by inorganic chemicals used in the west Kano irrigation scheme. An Unpublished MPhil Thesis, Moi University, Kenya.
2. Cairncross, s and Feachem R (1993). Environmental health Engineering in the tropics.John Wiley and Sons, Chinchester. Pgs 47-108
3. Nozaic D (2002). Water quality and quantity: in: Small water supplies: IRC Technical Paper series 40. IRC, The Netherlands. Pp 64.
4. Shaw, R (1999). Running water: More technical briefs on health, water and sanitation, pages 5, WEDC / IT, London

5. Skinner, B (2004) Water for low income communities. A WEDC postgraduate module. Part 1 page 1.6.; section 3.1.2.3; page 3.8. WEDC, Loughborough University.
6. Tayong, A (2002). Spring water tapping: in: Small Water Supplies: IRC Technical Paper series 40. IRC, the Netherlands. Pp153-160.
7. WEDC (2003). Community and management. WEDC Post-graduate module.Loughborough University.

Alternative responses / views to above questions Q1.1

- A slight mistake in your calculations – I make it 1.28 not 1.291 and 1.854 not 1.897 for the population growth factors.
- Although the 1995 census may have been reliable (and you could have discussed why rural censuses might not be) there have been demographic changes since that may have changed the growth rate. Use of the previous population growth rate (for 1985 – 1995) for 1995-2005 will mean that there is uncertainty about the reliability of any calculated value and a factor of safety would need to be built in to any designs unless further investigations take place into the actual population (and likely future growth).
- How did you conclude that AID deaths and immigration would 'balance out'. They would certainly have opposite effects on the population growth but may not balance out.
- You mention that the headman's figure may be an overestimate (yet it is less than what you calculated!). Did you mean he had made a mistake or that he had purposely given a wrong figure? You could have indicated why people may sometimes purposely give false population figures.
- You could have compared your projection for 25 years with that based on projecting from the headman's present figure for 15 years.
- It is hard to be sure of a reliable figure for the existing population based on the information provided. Therefore projecting for another 15 years is likely to be even more unreliable.

Alternative responses / views to above questions Q1.2 i)
The question was referring to whether different levels of accuracy are appropriate to each of the different systems more than to the design life of those systems (although that is relevant or to the feasibility of that source of water. You made some good comments about design life which were less relevant to the question posed. See model answer for my thoughts.

Alternative responses / views to above questions Q1.2 ii)

You mentioned some of the good points that I was wanting but missed others. See model answer for my thoughts.

Alternative responses / views to above questions Q 1.3

A very good answer. You could also have mentioned checking with other organisations in the area who may have already got some more recent data.

Alternative responses / views to above questions Q2.1

Your overall method looks is long the right lines but I query:

1. Your use for 45 l/p/day assumed for domestic use. Do you think that this is likely in view of the distance it will have to be carried. You could have checked typical walking distances. Drawing circles or radii 125m and 250m around the wells can give some impression of the distance people will have to walk.

2. Your allowance of the full water demand figure for livestock within the amount of water drawn from the wells. It would be much more likely that the cattle in particular would still be led to surface water. Would you like to lift 35 litres per head of cattle from a well if there was an alternative surface water source that you could lead the cattle too when they were taken to pasture. To raise the 5,000 litres from well W4 would take 4.2 hours of continuous work!

3. Your assumption of a boarding school in this rural village. I'd like you to have justified you estimate of the population of the school. What uses would you imagine 140 litres/head would be used for? It is not feasible for a hand-dug well with a bucket capable of drawing at only 20 l/minute to provide 63 m^3 per day! Also, how would used water be disposed of at the school?

Alternative responses / views to above questions Q2.2

You have not really concluded if the daily demand from the well will be affected. See model answer.

Alternative responses / views to above questions Q3.1

Unconfined aquifers can produce pathogen-free water if the soil is a fine granular soil and the depth of soil through which contaminants have to travel before reaching the groundwater is sufficient to ensure their removal. The time of travel in the groundwater is also relevant – longer times mean greater likelihood of die off.

You wrote about the general situation rather than focussing on what the map shows. You refer to septic tanks and sewers but these are unlikely in this rural area. You could have observed the houses above spring A and near to spring B and have mentioned the risk of pollution from any latrines serving these homes, particularly since the groundwater is very near to the surface at this point.

If the land above the spring is used for agriculture there may be a risk from fertilisers and pesticides.

Alternative responses / views to above questions Q3.2
You need to say more to get better marks. In particular discuss how storage allows the maximum potential of the spring to be used by ensuring no water overflows and how the storage allows a higher rate (i.e. to suit the peak demand rate) of withdrawal from the tank than the rate of spring flow into the tank The amount of water stored in the tank needs to be carefully checked at the design stage to ensure that, throughout the day, it can meet the pattern of demand without running dry and inconveniencing users of the water supply system.

Removing turbidity is best done in a separate tank that remains full (i.e. the water overflows from it into a storage tank after the solids have had time to settle). Hence sedimentation takes place throughout the night and day. If we use the storage tank for settlement the deposited solids will often be re-suspended when the water level in the tank becomes low and the flow of the water entering the tank disturbs them.

A fairly good critique of the pumped scheme. It would have been good to have mentioned how the wells with buckets and windlasses avoided some of the disadvantages of the pumped scheme.

It is good to use page numbers when you quote references (i.e. use Skinner (2004: page 1.6) in the text. They are not then used again in the list of references. Do not use the authors initials in the reference in the text.

RAIN WATER COLLECTION DESIGN
Essay: About a proposed rainwater collection system:
Assumptions:
- Roof surface coefficient for tile = 80% (WEDC, 2004 pg 3.10/ 3.12)
- There are no reservoir losses (leakages, evaporation) etc, i.e. top of reservoir is closed.

Water Supply:
The maximum Possible water Supply from the tile roof = 5m x 3.5m x 0.8 x 1300 mm pa = 18.2 m³/ annum.

Water Demand:

Daily demand = 95 l/day; = 95 x 365 days = 34.8 m^3/year. Therefore
(i) Given the annual total annual water demand (34.8 m^3) exceeds maximum possible annual water supply (18.2 m^3), the rainwater cannot be the only source of water for this situation. Also,
(ii) Given 20l/day is for essential water needs such as cooking and drinking, up to 20 l/day x 365 = 7.3 m^3 gives a reasonable minimum that should be available from rainwater for the whole year.

However, there are two completely dry months during which only rain water stored in the tank can serve. During this dry period, supply should outstrip demand- even if it means manipulating demand to satisfy this scenario. Thus Assuming 1 month has 365 / 12 days = 30.4 days, the amount of water required during the two dry months = 30.4 days / month x 2months x 95l/day = 5776 litres (or 5.8 m^3) (WEDC, 2004 pg 3.25)

Thus 5.8 litres of water will have to be used from the stored rain water during the two dry months. BUT the tank has only a capacity of 4.5 m^3, giving a deficit of 1.3 m^3. This means that the mandatory / essential 20l/day water demand for cooking and drinking can be extracted from the tank, giving a total essential demand of 1216 litres (1.22 m^3). A balance of 3.28 m^3 still in the tank can be distributed over the two dry months, but with daily water use of about 74 litres / day (instead of 95 l/day). This means that the other essential water needs of 0.02m^3/day can be met from the tank, while the non-essential water needs /uses per day reduce to at most 54 l/day instead of 74 l/day during the two dry months. For the ten wet months, the reservoir capacity is still too small, rendering conjunctive water use the only option. Up to 18.2 m^3/ 10 months = 1.82 m^3/ month is available (assuming uniform distribution of rainfall through the rainy season) (WEDC, 2004 pg 3.24-3.27). This gives about 0.06 m^3 / day possible water availability.

The extra rain water can still be collected from the tile roof using oil drums, plastic containers / barrels e.g. Skylak and other smaller containers. This can be coupled with the use of surface water in streams, ponds for other uses such as agricultural which does not need very pure rain water. This conjunctive use of water can help the community manage water needs well. Investment in tiles for roofing and reasonable sized water containers as a coping strategy may be tricky because of monetary demands- given marked socio-economic differences among households.

However, the hidden ability and willingness of most low income communities to invest in water supply coping strategies exists, and can support the situation here. WEDC (2003) pgs 109,112,122,133, 140-1; and WEDC 2004 pgs 2.4) state that the poor already pay more to access water. The key idea is to leave the management of the water issues to the community's lowest possible level / structure – i.e. subsidiarity- which promotes initiative and ownership- and hence sustainability of projects (WEDC 2004 b pgs 65) .

Essay 2:
A small rural remote community in a developing country use a caisson-lined open well in centre of village as its only source of water. Nearest alternative is a hand pump 1.5 km away. Groundwater 6 m below ground and water collected using bucket, rope and pulley. All use same bucket and rope which remain at well. Has concrete headwall 900 m high, with no apron slab but there are no obvious problems arising from it. If the results show that the fecal Coliform count is 40 E. coli bacteria per 100ml of the sample instead of the government's recommendation of 0 E.coli bacteria, what would you recommend is done and why? Ensure that any suggestions you make are likely to be feasible (one side of A4 sized paper); 15%

RESPONSE:
A caisson-lined open well is in the centre of village as its only source of water. The caisson lining is okay, but caisson joints may be loose, thereby allowing some of the contaminants from the soil, child faeces washings around the well, animals droppings, aqua privies, latrines etc to reach the intake (WEDC, 2004 b pgs 1.14, 4.9, 4.17 and Watt and Wood, 1979 pgs81-92). Caisson may also be expensive. The lining should be checked to confirm its water-tightness. If not, for the large gap between the caisson and parts of the shaft wall, the lower 3 metres should be filled with clean sand and gravel, while the top 3 m made water tight by applying puddled clay (WEDC, 2004b pgs 4.22; Watt and Wood, 1979 pgs 77; and Cairncross and Feachem 1993, pgs 1-15).

The open well further allows contamination from the soil, air and dying / dead animals (Watt and Wood, 1979 pgs 28-30; & WEDC, 2004 b pg 4.9). The nearest alternative is a hand pump 1.5 km away, which is not a feasible alternative as operation and maintenance (O&M) issues are likely to arise. Thus priority should be to make the well functional. Groundwater is 6m below ground, which is close enough to the sandy surface. Thus some E. coli may be gaining access to the water from the surroundings, and from the unlined wellhead. Since all use the same bucket and rope which remain at well, some users may be putting the bucket onto the ground, making surface contaminants getting into the water when in use (Shaw, 1999 pgs 25-28).

Causes of contamination, and proposed ways of addressing the same:

The most common sources of well contaminations are: well head, lining and water entering the intake (Watts and Wood, pg30). There should thus be an impervious apron to collect spilt water, and guide it to a soakaway. It is also useful to check that the shaft lining and the junction between the well head and the top of lining are water tight as they may be the sources of contamination. The top of the open well is a source of contamination (Watt and Wood, 1979 pgs 28-30 and 161), and should be covered by either a permanent or moveable well head or cover. A raised sealed cover may be installed, but may require a pump to access the water. The latter is not a recommended option due to logistics of O&M of the pump (Shaw, 1999 pgs 25-28). A movable cover may be most feasible in the short term as it immediately protects the shaft from windblown dust, insects, and other air-borne pathogens – some of which may be fecal (e.g. E. coli) when the well is not in use. It may be helpful to install an impervious apron 2 metres wide sloping away from the shaft in all directions, with a provision for apron drain / soakaway a safe distance away. This apron should be swept periodically (Shaw, 1999 pgs 25-28).

If well is new, it should have been disinfected before use to eliminate contaminants. It can then be periodically disinfected, especially if epidemic arises. The open well should be periodically drained, cleaned and disinfected during dry season (Shaw, 1999 pgs 25-28; & Watt and Wood, 1979 pgs 36, 99). There should be a clear site for washing, livestock drinking and bathing, and a fence around the well to prevent livestock coming close and defecating around it. This also saves the possibility of animals falling inside and dying in the well, leading to its contamination (Shaw, 1999 pgs 25-28). All latrines, aqua privies, etc should be more than 25 metres away from the well, to avoid contamination (Shaw, 1999 pgs 25-28; and Watt and Wood, 1979 pgs 61, 77, 99).

Essay 3: Soil at a particular site thought to be suitable for excavation of a hand dug well (1.3 m diameter internal) with a lining made of either burnt clay bricks or of reinforced in situ concrete. Describe 3 most important aspects considered most important to make one choose btn these 2 methods of lining(brickwork to be 225 mm thick; concrete lining 150 mm thick, each provided with a similar parapet wall, bucket and windlass, concrete apron slab and drain) (one side A4; 15 marks)

RESPONSE:
In general, reinforced concrete lining is the most appropriate, as it is easy to mix and lay, is relatively cheap in most places, and adaptable in other ways (Watt and Wood, 1979 pgs 25-27; and Shaw, 1999 pgs 25-28). Reinforced in situ concrete is poured in a special mould which has been lowered into the well and assembled in the already prepared cutting ring around the re-rod cage. Seepage holes are built into the casting using removable rods which are pulled out of the casting when it has set. The mould is stripped a few days after, and then additional rings are cast on top. It is also convenient because it does not need props, but may need curbs to support the concrete shaft every 5 m. These curbs are cast in place behind shutters made from either timber or a temporary brick walling and given at least one week to cure. This long wait is one disadvantage. It also uses the minimum of concrete (Shaw, 1999 pgs 25-28). It does not need a foundation because there is enough friction at the interface between the concrete and the soil to support the weight of the concrete (Watt and Wood, 1979 pgs 25-27; and 131).

Disadvantages:
However, the caisson tube tilts during casting, often trapping the external mould shutter. This requires that the caisson is then straightened up by excavating under the caisson and jacking the caisson vertical. This is difficult and tome consuming to correct. Filling the space between the caisson lining and the permanent lining with small round gravel helps to hold the caisson upright. The caisson is given several weeks to cure before it is sunken (Shaw, 1999 pgs 25-28; and Watt and Wood, 1979 pgs 25-26 and pgs 131).

Burnt Clay bricks:
This needs a foundation, which is wider than the normal excavation to support each section of the brickwork to avoid danger of part of the bottom of the lining slipping into the hole which will be excavated under it as the well is deepened. It also needs props- a clumsy process which makes the excavation and lining difficult. Brickwork also cannot be used in great depth or through unstable

ground where collapsing earth walls prevent the lining from sinking; The material are also very weak under tension, and the stress set up by collapsing earth sides can fracture the linings, especially if this occurs on one side of the shaft only (Shaw, 1999 pgs 25-28; and Watt and Wood, 1979 pgs 25-26, and 131).It is also difficult to make brickworks water-tight and prevent polluted surface water coming in. This makes it easy to contaminate. It also uses a lot of concrete to make the joints. Lastly, the brickworks involves extra works like filling the space between the lining and the excavation for the top 3 m with puddle clay or concrete mortar; while the inside of the well above the water table is also plastered to prevent growth of vegetation and insect life. These disadvantages generally render brickwork more difficult and inconvenient to use compared with the in situ concrete (Shaw, 1999 pgs 25-28; and Watt and Wood, 1979 pgs 25-26, and 131).

Essay 4: (Key reference: WEDC, 2005)
Spring tank usable storage capacity of 2.5 m^3 ; Flow into tank = 0.05 l/s ; No water drawn from tank between 2100 hrs and 6.00 am; Daily water demand = 4 m^3; 6 am – 9 am water demand = 2 m^3 and from 9 am – 9 pm = 2 m^3; Comment on whether or not the tank size and spring flow rate are suitable to meet the demands of the community.

Essay 4i:
Water flow rate = 0.05 l/s;
Water supply in 24 hours = 0.05 l/s x 24 hours x 6 x 60 = 4.74 m^3/day
Water demand over the 24 hours = 4 m^3.
Since water supply exceeds water demand, the spring is able to supply enough water to the population (Shaw, 1999 pgs5)
Demand rate during peak hours (6am – 9am) = 2 m^3 / 3 hours = 2000l/10800 sec = 0.185 l/s
This exceeds the supply rate (0.05l/s), and therefore there must be a tank to store the water overnight.
Water demand during peak hours = 2 m^3 while tank capacity = 2.5 m^3.
Therefore water can be reasonably supplied by the spring at peak time. If the tank is full at beginning of peak water demand, there would still be 0.5 m^3 of water left in the tank to start serving the medium water demand hours from 9 am – 9 pm.

Water demand rate from 9 am – 9 pm = 2 m3 / 12 hours = 2000litres/ 43200 sec = 0.04625 litres / sec
However, water supply rate = 0.05 l/s

Since this demand rate is lower than the spring water supply rate, there is enough to serve the rest of the day

Water supply overnight (from 2100 hours to 06.00 am) = 0.05 x 60 x 60 x 9 hours = 1.62 m³

Water supply from 6 am – 9 am = 0.05 x 60 x 60 x 3 hours = 0.54 m³

The scenario of water supply and demand, against the net water available in the tank is shown in the table below.

Assumption: Start with an empty tank at 2100 hours.

Table 1: Water supply and demand schedule between 2100 hours and 0900 hours

Time hrs	2100		0600		0900		2100		0600		0900	
Tank water level m³	0		1.62		0.16		0.32		1.94		0.48	
Water supply in m³		1.62		0.54		2.16		1.62		0.54		2.16
Water demand in m³		0		2		2		0		2		2
Net water in in m³		+1.62		-1.46		+0.16		+1.62		-1.46		+0.16

The table shows that even if the tank were to be empty at 2100 hours on any day, there would be enough water to serve the population, with each subsequent day leaving increasing amount of water in the tank. Therefore the tank size and the spring flow rate are enough to serve the demands of the community.

Essay 4ii:
Scenario 2: If the calculations were to start from 1700 hours when the tank is empty, and finish at 1700 hours the following day.

The calculations:
1700 hours to 2100 hours = 4 hours; water supply over this period = 4 x 60 x 60 x 0.05 = 0.72 m³

Water demand over the same period = 4/12 x 2 m3 = 0.67 m³. This gives net water availability at 2100 hours of 0.05 m³.

The tank is then left unused from 2100 hrs to 0600 hrs (i.e. a period of 9 hours). Net water supply over this period = 9 x 60 x 60 x .05 = 1.62 m³, which accumulates overnight, and giving a net water availability of 1.67 m³ in the tank at 0600 hrs.

From 0600 hrs to 0900 hrs, there is water supply of f 3 x 60 x 60 x 0.05 l/s = 0.54 m³. There is, however, a maximum / peak water demand of 2 m³, giving a net water accumulation of -1.46 m³ in the tank. This, plus the amount of water accumulated overnight of 1.67, gives a net water availability of 1.67-1.46 = 0.21 m³ in the tank by 0900 hrs.

Between 0900 hrs and 1700 hrs, the water supply is 0.05 x 60 x 60 x 8 = 1.44 m^3. However, there is water demand of 2/3 x2 m^3 = 1.33 m^3 of water used. The net water accumulation in the tank over the period is 1.44 – 1.33 = 0.11 m^3. This, added to the earlier water level in the tank of 0.21m^3, gives 0.32 m^3 of water in the tank at 1700 hours the following day. This completes the 24-hour cycle starting 1700 hrs.

The following table shows the demand-supply schedule describe above. Further details are in the attached graph (Figure 1).

Assumption: Start with an empty tank at 1700 hours.

Table 2: Water supply and demand schedule between 2100 hours and 0900 hours

Time hrs	1700		2100		0600		0900		1700
Tank water level m^3	0		0.05		1.67		0.21		0.32
Water supply in m^3		0.72		1.62		0.54		1.44	
Wate demand in m^3		0.67		0		2		1.33	
Net water in in m^3		+0.05		+1.62		-1.46		+0.11	

The table shows that even if the tank were to be empty at 1700 hours on any day, there would be enough water to serve the population, with each subsequent day leaving increasing amount of water in the tank. Therefore the tank size and the spring flow rate are enough to serve the demands of the community.

Essay 4iii.
Scenario 3: This scenario assumes that the calculations start at 1700 hours, with the tank having 1400 litres of water inside it. The following table shows the demand-supply schedule describe above. Further details are in the attached graph.

Assumption: Start with an empty tank at 1700 hours.

Table 3: Water supply and demand schedule between 2100 hours and 0900 hours

Time hrs	1700		2100		0600		0900		1700
Tank water level m^3	1.40		1.45		2.5**		1.04		1.15
Water supply in m^3		0.72		1.62		0.54		1.44	
Water demand in m^3		0.67		0		2		1.33	
Net water in in m^3		+0.05		+1.62		-1.46		+0.11	

143

** means that even though the total water quantity could be exactly 3.07 m3, the tank capacity is 2.5 m3. So the rest of the water drained from the tank overnight, keeping the tank full at 2.5 m3.

Essay 5: Prepare an outline design giving the critical plan dimensions only for each of the following stages of water treatment works which needs to be able to treat water at a constant rate of $5m^3/hr$ throughout the day and night every day of the year without any interruptions.

Sedimentation-assume that an acceptable surface loading rate is 0.25 m/hr
Slow sand filtration-assume that the maximum acceptable rate of filtration at any time is 0.15 m/hr   (5%)
According to WEDC, 2004 (pgs 6.22-3) and Smet and Wijk 2002 (chapter 12, 14, 15 and 17), Horizontal velocity of flow (v) = Q/BL where Q = Flow rate in $m^3/_{hr}$; B = width of tank; and L = length of tank.
In this case Q =5; v= 0.25 m/$_{hr}$ (or m^3/hr per m^2)

Surface loading rate = influent flow rate / surface area of tank. Thus v = surface loading, while BL = surface area of tank (Smet and Wijk 2002 (chapter 12, 14, 15 and 17.)
Therefore BL = Q / v = $5/_{0.25}$ = 20 m2.
According to WEDC, 2004 pgs 6.23-4, B:L should be in the range of 1:3 up to 1: 8
Using a figure of 5, 5B.B = 20; B = 2.
Therefore the length of tank = 10 m, while the width is 2m

According to WEDC, 2004, (pgs 6.23-4) and Smet and Wijk 2002 (pgs 315-316), Depth of tank should be adequate to accommodate the sludge accumulated at the bottom between the cleanings. A tank of 2 m or deeper could accommodate mechanical sludge removal equipment. Thus a figure of 2.5m is adopted for depth of tank.
If an allowance of 0.5 m is left as an allowance for sludge deposits before cleaning, horizontal flow velocity = $5_{/(2.5-2)}$ x 2 = $5/_4$ = 1.25 m/hr. This horizontal flow velocity is within recommended range of 1-3 m/hr (WEDC, 2004 pgs 6.24 and Smet and Wijk 2002, (pgs 316-317).

Slow sand filters:

Recommended rate of filtration = 0.1 – 0.2 (Quoted as 0.13 – 0.5 elsewhere) (WEDC, 2004 pgs 6.26-6.29; and Smet and Wijk 2002 (pgs333). Therefore the stated 0.15 is reasonable.

If $Q = 5 m^3/_{hr}$,

And maximum rate of filtration = 0.15 $m/_{hr}$, then

Surface area of slow sand filter = $5 /_{0.15 m^2}$ = 33.3 m. This is minimum, since WEDC, 2004 (pgs 6.28-6.29) and Smet and Wijk, 2002 (pgs 333) recommend a figure of less than 100 m^2.

The surface dimensions of the slow sand filter can be 3 m x 12 m.

(b) If in addition the water is being chlorinated

Decide at what rate (l/hr) a chlorine solution with strength of 1.5g cl/l needs to be added to the water flowing into a contact tank to give it an initial dose of 2mg/l

If 1.5 g Cl/L gives a dose of 2 mg Cl/L

Dilution rate = 1500 mg Cl / L: 2 mg Cl / L = 750: 1;

The chlorine Dosing ratio is 750 water: 1chlorine solution.

If desired treated water flow is 5 m^3/ hour, the chlorine solution flow (dosing) rate should result from $5 m^3/_{hr}$ x $1/_{750}$ = $^{20}/_3$ litres / hour.

Therefore the chlorine dosing rate = 6.7 litres / hour.

CHLORINATION TANK DESIGN

Essay: Draw in plan and show the critical plan dimensions for a suitable arrangement for a chlorine contact tank- assume that the water in the tank is 0.9 m deep and the minimum contact time required is 30 minutes (5%)

Essay 6:

Introduction:

Water treatment is necessary to remove impurities which reduce health and aesthetics of water (WEDC, 2004, pgs 6.4-6.6). Water may be treated at a central point, or at household (HH) level. The options (in order of preference) in water supply management are (i) have a good quality water source, and protect the source; (ii) treat water from a source with the highest quality; and (iii) maintain quality after being drawn from source. It is generally pointless removing pathogens if treated water is contaminated before use. There is thus a need to store water carefully in a covered clean container, and collect it hygienically. Our options here are (i) HH treatment or (ii) central treatment at source (pond); then use a hand pump in both options.

Option 1: The HH option is easier to manage, as responsibilities are clear, with each HH being in charge. They can choose how much to treat, treatment method, and for what use to put the treated water (WEDC, 2004, pgs 6.8-9). In this case, straining followed by slow sand filtration (SSF) is proposed. These are relatively easy and cheap methods (Smet and Wijk 2002 (pgs330-333). Straining helps remove the bulk of the macro pathogens such as copepods. SSF is able to slowly remove all other pathogens, especially after the biological layer matures. However, HHs need to be properly motivated to avail and clean the filters as required.

Socio-economic differences among HHs may render the universal household treatment tricky and unlikely. This can be partly ensured by regular health education (HE), as well as integrated development programs which also support income generation at HH level. HE can partly also help families aware of water storage and post-treatment storage and handling skills, as well as reduce contamination after treatment. It also gives each HH a leeway as to what kind of chemical to use, if any- as chemicals impart tastes which may not be universally acceptable in a community. HHs can choose what is acceptable to them. However, the other hand HH treatment of water is never a guarantee that it will be done by all, due to socio-economic differences. The clothes used for straining water at HH level may also be dirty, posing health risks (WEDC, 2004, pgs 6.6.20-6.21). Similarly, the diseconomies of scale associated with small scale treatment at household level may make costs of doing it prohibitive. This may demand extra motivation for water treatment at HH level, rendering the central treatment better.

Option 2: Central treatment (CT) of water: CT is a good option because it guarantees uniform standard for all HHS who rely on the same water source. It also comes with the advantage of economies of scale, which may make treatment cheaper per HH. However, the issues of O&M, coupled with acceptability of taste of treated water (Smet and Wijk 2002 (pgs 418), among other issues are likely to arise in commonly managed water sources. Community members must be willing to cooperate in O&M issues of a communal water treatment facility. Distance from HHs may also render it useless for some HHs, which may resort to nearer, and perhaps worse quality water sources.

Related to this is the possibility of tragedy of the commons ensuing, in cases like SSF cleaning whereby nobody may take trouble to clean it (WEDC, 2004 pgs 6.9, 6.22, 6.26). In this respect, responsibilities need to be clear. The use of a pump is particularly tricky, as it is recommended that however simple an innovation involving a pump looks, an option of avoiding pumps should be first sought. Due to common stakes most have in the system, however, it is likely to be cared for well, with all ensuring source contamination is avoided. Both options involve use of a pump; any involving motorized pump should be avoided.

One clear disadvantage of CT is that it may reduce the number of people getting water because the two taps may not be sufficient at peak hours, leading to scrambling, and general disillusionment. A related concern is the possibility of not having enough storage for treated water after SSF (WEDC, 2004 pgs 6.26-29). SSF may give good quality water but its speed of operation may be overwhelmed by the community water demand. This is likely to make others resort to alternative water sources, some of which may be worse than what the communal treatment point provides. Secondly, there may be cultural issues associated with change of water tapping method- stepping into the pond, and scooping water directly- as opposed to the proposed new system of drawing what is clearly pond water at a different point. Thirdly, a difficulty of getting water during maturation of biological layer and cleaning of SSF arises.

According to Watts and Wood 1979 (pgs 28-30, 34-38), the quality of water supplied at a water point does not necessarily relate to the water quality which is eventually drunk. Good quality water at rural collection points become contaminated before consumption. Health education initiatives may help reduce this risk, by encouraging the adoption of hygiene water handling practices (Smet and Wijk 2002 (pgs 67, 265).

Water is transported in open containers with branches of shrubs and trees added to stop water sloshing about and spilling (WEDC, 2004). This, among others, are some of the possible risks of contaminating water after treatment from the CT point – at the stages of collection, handling, transporting, home storage and handling at home. This may be worse in HHs which have total confidence in the CT so that they do not bother to take precautions. What is useful is agreement by all users on O&M arrangements to maintain water quality (IRC 1995, pgs 45)

Whichever way, there may be need for health education program to help the HHS to hygienically handle and store water after treatment whether at home or at the CT. Commonly, insects and rats die in water systems, while dirty containers are used to scoop water from the HH storage containers (Watt and Wood, 1979, pgs 36-38).

REFERENCES:
1. Cairncross, s and Feachem R (1993). Environmental health engineering in the tropics. John Wiley and Sons, Chinchester.
2. IRC (1995). International water and sanitation centre, The Hague, The Netherlands
3. Shaw, R (ed)(1999). Running water: More technical briefs on health, water and sanitation, WEDC / IT, London
4. Skinner, B (2004) Water for low income communities. A WEDC postgraduate module. WEDC, Loughborough University.
5. Smet J and Wijk C (eds)(2002). Small Community Water supplies: IRC Technical Paper series 40. IRC, the Netherlands.
6. Watt S and Wood W, 1979 Hand dug wells and their construction. ITDG, UK.
7. WEDC (2003). Community and management. WEDC Post-graduate module. Loughborough University.
8. WEDC (2004).Water for Low income communities-part 1. A WEDC Post-graduate module. Loughborough University.
9. WEDC (2004b). Integrated Water resources Management (IWRM). A WEDC Post-graduate module. Loughborough University.
10. WEDC (2005).Water for Low income communities-Topic task 2. A WEDC Post-graduate topic task for April. Loughborough University.

WLIC Topic task 2 Feedback
Critique of the responses to questions above
A good answer.

At the start of the third paragraph I think you mean 5.8m^3 not 5.8 litres! After writing about the reduced demand of 74 l/d you say that 'the reservoir capacity' is too small for the ten wet months. Is it not rather a problem of shortage of rain to provide more than the 60l/day average supply?

It is not clear for which scenario (20 l/d or 60l/d or 95l/d) you are referring to when you speak about the 'extra rain'. Are you referring to providing additional catchment area?

Q2 : A very good answer with regard to appropriate well-protection measures. Good use of references.

In addition I would like you to have mention the following:

▪ It is worth retesting the well water to see if the original test is a result representative of the normal situation and not just a one-off occurrence.

▪ It is also worth testing the quality of water that people consume at their houses since even if the well water has zero e.coli it may be contaminated before consumption. Efforts to improve the well water will be wasted if it is always being contaminated after collection!

- Health and hygiene education are to be recommended to preserve the quality of the water and to help people avoid other transmission routes of water-borne diseases.

Q3 : The presentation was a challenge; should have been presented in the form of a table and expected three clear aspects (e.g. 'ease of construction', 'strength of the lining', 'amount of specialised equipment required' etc) that you would consider when making a choice between the two methods. Gave too much general information about both lining methods without clearly comparing them under just three aspects.

You speak of the in-situ method using the 'minimum of concrete' yet surely it uses more concrete than a brick lining? For the brick lining you say 'uses a lot of concrete to make the joints' yet in fact the joints are made with cement mortar. I doubt if more cement is used for the brick lining than for the in-situ concrete but because of the different thickness of the linings and the higher cement content of the mortar I suppose that it might be possible.

You mix up in-situ lining with use of caissons. Caissons are normally pre-cast outside of the well whereas in-situ lining is cast in the well directly against the soil using only an internal shutter.

Incidentally, note that reinforced concrete curbs can be used with brickwork lining to avoid the use of the more complicated brick curbs when successively using 'dig a portion then line a portion method'. These concrete curbs can often reduce, or eliminate the need for props. If long sections of excavation can be kept safe with temporary supports (e.g. the modified Chicago method) it is possible to line from the bottom up which eliminates the need for props when using brickwork.

Q4
You made a calculation mistake on your second line. It should be 4.32 not 4.74 but fortunately it did not affect the rest of your calculations. Otherwise very good work for parts i) and ii). It would be good to label with amounts the volumes in the tank at important points on both your graphs.

Your graph for part ii) is incorrect in that the cumulative supply line should be 1400 higher and should be level between just before 02:00 hours (1:49 to be exact) and 0:600 hours before continuing at the same gradient. This is because during this time period the tank remains full and water overflows. Consequently there should also be a flat peak to your 'water quantity in tank line' for the same period at 2500 litres. Your table is right but you did not calculate information for some of the important intermediate points.

Q5

Your sedimentation tank calculations are correct for one tank, but you have not made provision for cleaning or maintenance. Using a number of parallel routes of lower capacity, so when one is out of commission the total flow through the others does not exceed a loading rate of 0.25 m/hr is advisable.

Your area for the SSF is correct but again what will happen when you need to clean it? The flow will have to be interrupted which is unacceptable. Using say four filters in parallel (each 11.1m^2) reduces the additional provision required from 33m^2 to 11.1 m^2 and this will normally save cost.

Your dose rate is correct but you did not provide a diagram for the layout of the chlorine contact tank.

Q6

A very good answer that covered a wide range of technical and non-technical aspects. You exceeded the one side limit but wasted space repeating aspects, so I have not penalised you. Good use of references.

One important aspect you did not mention is the level of training (e.g. cleaning and eventually replacing the sand) and follow-up needed to ensure every household uses and maintains the HH filter properly. Also note that treatment at household levels allows treatment of only that water that will be used for potable purposes.

WATER FOR LOW INCOME COMMUNITIES TOPIC 3:
Question 1:

Carry out the necessary calculation so that you can enter the missing values in each of the highlighted boxes (i.e., those with a thick border line) in tables 3 and 4. On a separate piece of paper clearly explain the method you use to calculate the missing figures needed in table 3 for pipes IJ, CH, BC and AB. You are not required to show your method for any of the other figures.

CALCULATIONS FOR TABLES 3 AND 4 (WEDC, 2005 pp 3.4-3.7):
Calculations in this part of the question are done as per the following references: WEDC,2005 pg 3.3-3.8; WEDC, 2004 b pg 9.18; Smet and Wijk 2002 pp 442 -446 ; Jordan, 1980 pp 70-77; and WEDC, 2005 pp 3.4-3.7
IJ: Using WEDC Additional resources WAVIN Chart:
According to the waving chart, head loss factor when pipe Diameter = 25.75, and at a flow rate of 0.3 l/s = 0.02 (i.e., 20m/1000m). Since distance IJ = 95 km, the total head loss = 0.02

m/km X 95 km = 1.90 m. But since the design figure cannot be less than the calculated value, it is rounded up to 2.0 m.

CH: Required elevation of HGL (m) ASL for pipe CH = Required elevation of HGL (m) ASL for stand post + actual head loss (rounded up) (m)
Thus required elevation of HGL (m) ASL for pipe CH =283.5 m + 0.3 = 283.8m. This becomes a control point.

BC: Required elevation of HGL (m) ASL for start of branch 2 from C / Pipe BC = Control value at the start of branch 2 at C (written above, as control point / value) = 283.8m.
Required elevation of HGL (m) ASL for Point B start of branch 1 of Pipe BC = Required elevation of HGL (m) ASL for start of branch 2 at point C + actual head loss (rounded up) for pipe BC (m) . Thus required elevation of HGL (m) ASL for start of branch 1 at B = 283.8 m + 0.5m = 284.3m.

AB: Required elevation of HGL (m) ASL for start of branch 1 at B = the figure at the control point immediately above at B for pipe BM = 282.0m

Required elevation of HGL (m) ASL for tank A = Required elevation of HGL (m) ASL for start of branch 1 at B + actual head loss for pipe AB (rounded up) (m) = 282.0 + 0.4 m = 282.4m.
Also show the calculations necessary to check if it is possible to reduce the diameter of pipe BM to 25.75 mm (internally) and still obtain at least 7 m residual head above ground level at M.

Pipe BM Characteristics: (WEDC (2005) pg3.3-3.8)
Peak flow rate = 0.27 l/s; Internal diameter = 25.75 mm; Pipe length = 110 m.
Head loss factor = 15/1000 (i.e., 15 m / km)
Actual head loss =   .015 x 110km = 1.65 m.
Residual head = 285.7 -1.65-27 9 =   5.05
Therefore the reduction of Pipe BM to diameter 25.75 would not avail the mandatory minimum head of 7 m. It would instead provide a head of only 5m- which is too low, and thus the change is not recommended. (15%)

Essay 2: Estimate if the figure shown in table 1 for the peak demand at standpost L is a reasonable figure for design purposes. There are 90 day pupils and 5 staff at the school to be served by one 20mm bib tap on a stand post. The tap is 1 m above ground level.

Show and clearly explain your calculations and any assumptions that you make.

RESPONSE:
90 day pupils @30 litres / person / day = 2700 litres / day
5 staff @ 40 litres / head = 200 litres / day (WEDC 2004b pg 8.6; WEDC 2004a pp 1.12/3)
3 households / huts getting water from tap L, with each household having 5 people = 15 people @ using 30 litres / day = 450 litres / day.

Total daily water demand at tap L = 3350 litres (of which 2900 litres is used by the school community the whole day, but 75% of this is used between 12.30 pm and 2 pm; i.e. 1.5 hours);
Thus 2500 litres / 1.5 hours = 2200 / 1.5x60 x 60 l/s = 2200 / 5300 l/s = 0.41 l/s
The figure given, however = 0.36 l/s

These two are not very significantly different, given they are both a result of assumptions which may be all realistic.
With 20 mm pipe diameter, and a peak demand of 0.41 l/s, the figures look like not matching, as it is not possible to have a 20mm pipe with a peak demand of 0.36, not even 0.41. So whereas they are both realistic figures, the tap diameter (20 mm) is out of range, and only taps of diameters such as 50 mm, 40.4 mm, 25.75mm, or 20.25 mm is possible (WAVIN Chart, WEDC 2004b pp 9.18). 20 mm is too low for the said flow rate, as the highest flow rate it can manage is less than 0.2 l/sec.

Assumptions:
The staff also operates from their homesteads and do not love in the school;
Other community members use the tap L during and after school- i.e. it serves both the school and the surrounding homesteads close by.
75% of the water is used at peak of 1.5 hours per day. The rest of the time would be water use by the community around.

Essay 3: When the tank is full the head available at is 288.5 m ASL (2.8 m higher than the minimum of 285.7m ASL assumed in the calculations). What would the general consequences be for the behaviour of the piped system whilst this additional head is available at the tank? (5%)

ATTEMPT ON THE QUESTION:
As long as the tank is full, the top water level is the automatic head.
. However, as the level falls, the design for the dynamic flow would
be based on the lowest water level expected in the tank. This is to
ensure that water can still be delivered at sufficient rate to the
distribution system even when the pressure at the tank is minimum
(WEDC, 2004b pp 9.9). AS the pressure is maximum, when the taps
are first opened, and enters the piped system, it will fill it and rise
up in each of the tubes until it reaches a horizontal line which will
be the same level as the surface of the water in the tank. If there
were an overflow system to the tank it would even reach there. –
i.e. the static level. Thus in this case the static level is 288.5 m. The
datum is the 285.7 m level. WEDC supports this, as the static level
is above the datum (WEDC, 2004b pp 9.9). The 2.8 m is the static
head / pressure

Essay 4:
These figures imply that:
Head loss factor == 4.75 m/1000m;
If pipe = 340 m, actual head loss = 0.00475 x 340 m = 1.615 m = 1.7
m (rounded up)
288.7 - 266. – 1.7 = 21 m

Pump: Rotates at 34680 rpm;
The total daily demand (including losses) at the end of the 15 year
design life of the scheme is 50 m^3/day.
This boils down to 50,000 litres / 24 x 60 x 60 sec = 0.5787 l/s
Population growth rate is 3% per annum;

(a) Carry out appropriate calculations and plot the results on a
copy of graph 1 (which should return with your work) to show
which of the four alternative impeller diameters is best suited to the
scheme. Justify your choice of impeller. using the chosen impeller,
at what flow rate will the pump discharge water into the tank and
what is the pressure at the pump (in N/mm2) if the centerline of the
pump is at a level of 262 m ASL?

Attempt on Essay 4a:
The calculations are done and results shown in the table below.
These are further used for plotting the graph below (separate
sheets-graph 1). The basic figures used in the calculations are first
read from graph 1 which shows flow rate, head and efficiency. The
calculated power is according to equation Power in Kilowatts = g x
Q x H / n in WEDC, 2004b pp 11.6), where g =9.81 m/s^2; Q = flow rate

in m³/s; and H is the pumping head; and n = efficiency (in decimal). To get the power adjustment figure (3.5), the average of all figures used in pumps of diameter 115 mm and that of 140 mm were used. The average got was 3.5, which was used to adjust all the other power figures fir pumps of diameter 120 mm and 130 mm. This adjustment was necessary to enable them fit in the scale given in graph 1. Accordingly, the figures have been used to plot the second graph in graph 1, 2005.

Going by the results as can be seen in table 1 below, Impeller of diameter 140 mm seems the best because it can provide water up to a head of 26.5 (the highest), with the highest level of efficiency of 56%. All the others can only produce lower heads, and still with lower maximum efficiency levels. It is also the only impeller with a capacity of delivering a head of more than 26 m. It is for this reason that impeller with a diameter of 140 mm is considered best suited to the scheme. The calculated figures are as shown in table 1 below.

Given the pump centerline is at 262m, while the tank elevation is 288.7, the head = 288.7 -262 = 26.7 m
According to Jordan (1980) pp 36, 1 m head = 1 kg/cm² water pressure. According to WEDC, 2004 b pp 9.7, 1 metre head of water = 0.981 N/mm². Therefore 26.7 m head. At this head, the flow rate (as read from the graph 1) = flow rate at maximum efficiency level (56%) = 20.2 m³/hr.

The equivalent pressure at this head = 26.7 kg/cm² = 2.67 kgf/cm² = 26.7 x 0.981 N/mm² (WEDC, 2004b pp 9.7). Thus the pressure at the pump = 26.19 N/mm² (About 26.2 N/mm²). However, considering the efficiency level of, the actual pressure = 26.2m x 56% efficiency level = 14.67 N/mm²
Given the facts in the graph and the assumptions, the following can be inferred:

Table 1: The characteristics of the different impellers showing head, flow rate, efficiency and power.

Impeller diameter	Head (H) (m)	Flow rate (Q) (m³/hr)	Efficiency (E) (%)	Calculated Power (CP) (KW)	Recorded power (from graph 1)	Dividing/ adjusting Factor	Adjusted power KW/3.5
115 mm	12.5	20	48.5	5.050	1.65	3.1	
	15	17.5	49.6	5.076			
	17	15.5	49.7	5.201			
	17.5	14.8	49.7	5.113	1.5	3.4	

	20	11.5	46	6.173	1.3	4.8	
140 mm	20	26	50	10.202	2.8	3.6	
	22.5	24	53	6.290	2.7	2.3	
	25	22	55.2	9.780			
	26.5	20	56	9.260	2.6	3.5	
	30	17	53.5	9.350	2.5	3.8	
	32.5	13	49	8.460	2.3	3.6	
130 mm	17.5	22.5	50.5	7.650			2.19
	20	21	53	7.770			2.22
	23	18	54	7.690			2.2
	25	15.5	53	7.170			2.05
	30	6	30	5.700			1.63
120 mm	15	20.5	49.7	6.070			1.74
	17.5	18	50.7	4.840			1.38
	19	16	51	5.840			1.67
	20	15	50.7	5.820			1.66
	22.5	12	47	5.640			1.61
Averag						3.5	

NB: In each case, power = 9.81 x H x Q x 100/ E

Table 2: Characteristics of the 4 impellers:

Propeller code	1	2	3	4
Diameter (mm)	140	130	120	115
Head at maximum efficiency level (H)(m)	26.5	23	19	17
Flow rate (Q) m^3/hr	20	18	16	15.5
Highest efficiency level attained (E) (%)	56	54	51	49.7
Power at highest efficiency point KW***	9.26	7.52	5.85	5.27

***NB: Power = 9.81xQxHx100/E (WEDC, 2004 b pp 11.6)

(b) Bearing in mind the information provided about the scheme at the start of this question, comment on any important implication that would result from using the specific pumping rate you calculated in part a of this question for the design. Also explain how you would prevent the tank overflowing if it becomes full. Limit your comments to one side of A4-sided paper.

ATTEMPT ON Essay 4b:
Therefore 26.7 m head. At this head, the flow rate (as read from the graph 1) = flow rate at maximum efficiency level (56%) = 20.2 m^3/hr.
The equivalent pressure at this head = 26.7 kg/cm^2 = 2.67 kgf/cm^2 = 26.7 x 0.981 N/mm^2 (WEDC, 2004b pg 9.7). Thus the pressure at the pump = 26.19 N/mm^2 (About 26.2 N/mm^2).

155

However, considering the efficiency level of, the actual pressure = 26.2m x 56% efficiency level = 14.67 N/mm^2

ATTEMPT ON Essay 6:

The following will need to be addressed as the plans to introduce Afridev are made. They largely cover preparatory stage, implementation and operation stage, and monitoring and evaluation phases. They are discussed by subtopic below.

Review of Past Afridev Pump performance records and development:

Review and analysis of the performance of the available Afridev pumps (pilot project) in the area is a first useful step (WEDC 2004b pp 7.60-7.61 and 12.1-12.5) to undertake. This can identify strengths, weaknesses /problems, threats/constraints, and opportunities related to the Afridev organisation, finance, personnel, logistics and technology. This can help to positively judge and sell it to the potential clients and stakeholders. If its past performance records look good, it can then go through other formalities (IRC, 1995-pp 53-55).It can be used to establish guiding principles and decode on the best approach to be promoted (IRC, 1995 pp 55).

Partnership:

There is need for partnership among different stakeholder-the Government, private sector, potential pump users, and local artisans- so that areas of misunderstanding can be addressed. It will also give confidence and security to those likely to be impacted by this scheme, and can leave the other pumps they may have been dealing in and adopt the Afridev pump. A full stakeholder analysis will be useful as a first step towards this partnership (Shaw 1999 pp1-4); WEDC 2004b pp 7.60-7.62 and 12.1-12.8; WEDC, 2003 pp 36-40). This can also help identify stakeholder roles, responsibilities, interests; stakes etc- and can ascertain feasibility.

Skills:

Implementation of the scheme will need skills training. A training needs assessment would be done, focusing on the local levels- with the VLOM as the ultimate goal. It is not clear who work in the local factory (they are likely to be outsiders- even expatriate). Thus there would be need for capacity building at local levels- VLOM approach so that 1-tier O&M would be established as the ultimate target even for those starting from 2-tier system (IRC,1995pp 29).

Tools:

Since some Afridev tools e.g. spanners are too specialized, there would be need for local manufacture of a variant of these (IRC, 1995 pp 29; Shaw 1999 pp1-4).). If it were in Kenya, I would involve the Jua kali artisans, in collaboration with the village and national polytechnics and technical training institutes in designing local tools for use. This will aim at reducing the monopoly likely to be associated with the single manufacturer in the country- thereby ensuring sustainability.

Standardization:

I will recommend the pump technology to be standardized- to make it familiar to the community, users and maintenance personnel. This is to ensure that (i) maintenance personnel become familiar with the equipment, spares and tools (ii) Users become familiar with the best way to operate equipment for long life; (iii) Standardize training of maintenance personnel (IRC, 1995 pp 29; Shaw 1999 pp1-4).). Preventive maintenance will be emphasized. The process should encourage exchange of Afridev parts with the other equipment which have been in market for some time- i.e. guaranteeing some level of compatibility (WEDC 2004b pp 7.60).

Policy on donation of equipment:

I will recommend a policy on donation so that only useful and durable parts of proven quality are be accepted. This is because there may be overdependence on low quality donations- which may render the program reliant on just one more expensive and unreliable source likely to be associated with too much solid waste to dispose of. Diversification of sources of parts and equipment will be recommended so that the best (with the highest comparative advantage) survive the competition (WEDC 2004b pp 7.60-7.62 and 12.1-12.8)

Coordinating body:

I will recommend the establishment of a well structured coordination body for the project. This can partly help establish standards, streamline import policies, establish a spare parts distribution system, operating and stocking licenses, material substitution, inspection procedure, training and hiring local artisans to make alternative spare parts for the same; holding of stocks (i.e. managing a national stocks store); and spare parts monitoring and feedback etc (pp 30-33; Shaw (1999) pp1-4; WEDC 2004b pp 7.60-7.62 and 12.1-12.4)

Spare parts:
The issue of spare parts will be first investigated, established and assured before deciding on the scheme (WEDC 2004b pp 7.60and 12.1). Spare parts will be ordered and stocked in bulk- and shared between communities. Maintenance tools will be standardized and limited to basic essentials, with the private sector involved in their distribution, stocking and sale. Material substitution will be emphasized- so that local initiatives which can manufacture high quality parts compatible with the pumps are encouraged to do so. There would thus be need to train the private sector- and other partners on the pumps- and enlist their support in stocking and marketing (IRC 1995-pp 28-29 and 30; and Shaw (1999) pp1-4).). There will be a need to register local businessmen to stock the spare parts, so that this information can be shared with the consumers. The government will be asked to intervene to guarantee availability of the spare parts so that the project takes root- perhaps by providing capital for materials, labour, and overhead costs and storage at the beginning to kick-start the project. Once there is enough demand, the local stockers can sustain the supply.

Subsidized parts:
It would be useful to avail subsidized spare parts to encourage consumers to adopt the Afridev technology. This is partly because the other technologies/ pump models are likely to fight back- and sabotage the whole process- e.g. possibility of bulk purchase and hoarding of spare parts by competitors- so that users are discouraged by unavailability and poor supply of parts- and thus maintain the status quo (i.e., the old models). This can be partly solved by stakeholder analysis, and keeping records of who has bought what (WEDC 2004b pp 7.60-7.62 and 12.1-12.8).

Ownership and O & M responsibilities:
We will define and clarify of who owns the scheme, and who has responsibility over O & M (IRC, 1995 pp 29).The scheme will clearly define who is in charge of O & M activities, - and who manages skills training, budgeting, revenue collection, basic accounts, careful monitoring of the treatment processes by an independent agency (IRC, 1995 pp 29; Shaw (1999) pp1-4).). As much as possible, the community will take this responsibility. A

record of who has bought what parts will need to be kept (WEDC 2004b pp 7.61-7.62 and 12.6-8).

Cost of project:
This will be clear, including who pays the capital cost (users should pay- to ensure sustainability); who pays the Operation and Maintenance expenses; and methods of funding future incremental capacity required.

Monitoring and evaluation:
There is need to keep records of the Afridev project as a means of monitoring progress. Record review, analysis, and sharing among stakeholders will be encouraged so that grey areas can be taken care of. It would be useful to monitor and evaluate the technology; community management; gender balance; agency support; spare parts production, distribution and sale; training needs; and training programs for those involved (Shaw, 1999 pp1-4; and WEDC 2004b pp 7.60-7.61 and 12.2-12.4).

When all these, among others are addressed, sustainability of the scheme is most likely to be guaranteed.

REFERENCES:
1. Shaw, R (ed)(1999). Running water: More technical briefs on health, water and sanitation, WEDC / IT, London
2. Smet J and Wijk C (eds)(2002). Small Community Water supplies: IRC Technical Paper series 40. IRC, the Netherlands.
3. WEDC (2003). Community and management. WEDC Post-graduate module. Loughborough University.
4. WEDC (2004a)Water for Low income communities-part 1. A WEDC Post-graduate module. Loughborough University.
5. WEDC (2004b).Water for Low income communities-part 1. A WEDC Post-graduate module. Loughborough University.
6. WEDC (2005).Water for Low income communities-Topic task 3. A WEDC Post-graduate topic task for April. Loughborough University.

SECTION 5: INTEGRATED WATER RESOURCES MANAGEMENT (IWRM)

Essay 2: Question 1a: The importance of social, human, environmental and financial factors in managing water quality.

Introduction:
A lot of factors interact with others to affect water quality, and are useful interventions in water quality management. They include economic, political / policy, physical, environmental, human, social and financial factors, among others. In this situation, only the last four factors will be discussed in as far as they are important in water quality management.

Social factor:
The society norms, values and organizational structures have been very useful in water quality management (WEDC, 2003). Traditionally, there were water resources that were never to be stepped on when collecting- as a society rule, and heavy penalty was meted against any defiance. There were water resources for different purposes – e.g. some ponds were specifically for livestock and general, others specifically for human drinking. There would normally be village guards to ensure the rules were adhered to. Bathing, urination and defecating were restricted to some sides of the water points to minimize water contamination and maintaining water quality. The society also ensured that only adults–went to fetch water in sensitive points such as drinking water ponds. This was to ensure that in case there was need for water resources maintenance e.g. manual dredging during dry season, the adults would make their rightful contribution. In some cases, some water resources were specifically for religious purposes, and could only be used under extreme circumstances of drought, and even then only under strict control. There were even some trees never to be cut- and children were made to believe cutting them was associated with a bad omen. It was an effort meant to conserve water, soil and biological resources. There was therefore always water of the right quality for each activity, including for drinking even at the driest month (personal experience 1970's-1980's).

The contemporary societies is quite heterogeneous partly due to migration, education, cultural changes etc. Ince and Howard (1999) state that the basis of most water quality standards is the WHO guidelines for drinking water. This can only provide guidelines, but enforcing them in a third world country need a lot more. Education of the society can help them understand the interactions among agriculture-industry- water and soil quality Vs health. This is because key social aspects such as culture, education, gender and health issues have become increasingly complex. Thus at the moment, the society needs to effectively contribute in enforcing water quality issues. The society, being made up of farmers, need to understand why prohibiting farming in some parts of a catchment is useful; not allowing people to step in water when collecting; using the same bucket for drawing water from a well; prohibiting waste disposal in some direction of and within some distance from a water point e.g. a spring (Jones, 1997; WEDC, 2004; personal experience)

Environmental factors

Environmental factors which could contribute to water quality include: rock type, wind, rainfall, temperature, soil type, evaporation, among others. High temperatures may increase evaporation which, in turn, may increase concentration of solutes (Total dissolved solids or TDS) in water. This may render some components poisonous, as their threshold levels will have reached. Similarly, when there is low rainfall, high temperatures and high evaporation, dilution ability of water courses such as rives is much reduced. Thus industries which may have been discharging its effluents into the river earlier, may all of a sudden find that it now pollutes. Strong winds have a similar effect as the factors discussed in this paragraph (WEDC, 2003; Afullo, 2003; Jones, 1997; and WEDC, 2004).

Similarly, erosive rainfall is of high intensity (more than 25 mm/hr) (Afullo, 2003; Jones, 1997 and WEDC, 2004) and may cause increased total suspended solid (TSS) loads in surface waters. Some soil types are also easily degraded, and if coupled with erosive rains common in the tropics, often lead to highly turbid waters high in TSS. Sometimes, storms may lead to flooding, leading to mixing of toilet contents with surface water. This tremendously contaminates all the available water, leading to reduced water quality. On a similar vein, some rocks comprise soluble salts. These would dissolve as water passes through them, leading to high TDS which affect taste and colour of water (i.e. physical water quality). Some of the rocks also have high buffering capacity, especially the ca and Mg-rich rocks. These stabilize the pH of water as it flows through them, thereby greatly modifying its chemical quality. This is particularly significant in acid-rain prone areas, as the resulting acid rain water is neutralized and stabilized.

Lastly, the catchment hydrogeology is useful determinant of water quality (Afullo, 1995). Flatter areas give water time to settle, thereby increasing residence time and purification occurs. The reverse is true for sloppier environment (WEDC, 2004; Afullo, 1995; WEDC, 2003).

Human factors:

These cover aspects of modernity associated with man. The main human activities include industrialization, urbanization and infrastructural development, agriculture, mining, education, technology and training, improved nutrition and health, among others. These have played significant roles in water quality management. Industrialisation, agriculture and urbanization have led to increased water pollution by heavy metals, pesticides, organic solvents, acidic / basic substances, aromatic compounds etc. This has led to reduced water quality, and there have been more common cases of water-related diseases such as water-borne, water washed, water-based and insect vector borne diseases, including chemical poisoning. However, education has increased awareness about these effects on water quality, and has subsequently led to increased focus on capacity building. Capacity has been built more on landfill (to reduce leachate pollution), afforestation and reforestation to reduce TSS; water works to remove water pollutants before use; effluent treatment (and pre-treatment) facilities to reduce pollution and contamination of water bodies; etc. Roof catchment is also becoming a popular water management tool meant to avail water of reasonable quality before it is contaminated by ground (soil and rock) based pollutants. Similarly, farm-based human interventions such as terracing, bunding, contouring, stabilization structures have been used to reduce soil particles compromising water quality and quantity (i.e., are used as water and soil conservation structures) (WEDC, 2003; Afullo, 1995; Afullo, 2004 and WEDC, 2004).

Financial factors:

Water quality depends on financial resources. Finances can be used to construct and / or purchase physical structures for soil and water conservation such as terraces, bunds, dams, gabions, waterways, etc. Finances are also useful in purchasing equipment for constructing water conserving structures above- including bulldozers, tractors (for dredging and making dams); JCB for making effluent and solid waste management facilities, etc. Man also uses finances to construct roof water catchment facilities, which significantly improve availability of good quality water to households, besides conserving soils and reducing siltation. Financial factors relevant to water quality are the capital and running costs for water facilities and services. This depends heavily on availability of credit facilities through formal and informal institutions (Afullo, 2004; WEDC, 2003; and Afullo, 1995).

Essay 1b: Discuss: water quality is best improved by increased hygiene education rather than treatment.

Generally, poor water quality is a threat to water resources, wildlife, and even man (Durham and Jackman, 2001; Ince and Howard, 1999). Most pollutants are from agriculture, industry, domestic / municipal etc (Afullo, 2003; Jones, 1997; Afullo, 1995). There are various ways of improving water quality, including treatment and otherwise. Water treatment normally involves physical separation of bulky solids, sedimentation, chemically assisted sedimentation (use of alum and water conditioners such as lime to correct pH), and lastly, biological treatment to eliminate pathogens and other biological agents. Whereas this makes the water safe, it is sometimes not palatable, with offending taste, smell, texture, look and colour (i.e., poor organoleptic acceptability) (WEDC, 2003)

Education is a tool for changing peoples' knowledge, skills, attitudes and practices in a given aspect (Afullo, 2004) . It is meant to improve their life skills, and is thus a vital aspect of capacity building. Hygiene education therefore means a deliberate effort to refocus peoples' attitudes, knowledge and practices in aspects touching on cleanliness of their surroundings, including their own bodies. Personal and environmental hygiene are most important in this context. Environmental hygiene education is meant to make people know the hazards around them, and how they interact with other materials to, for instance, make water impure (Afullo, 2004). It is a form of moral persuasion, which has the advantage of being sustainable. It involves both theoretical and practical aspects of water quality, which permanently impart a change to a society, as convinced parents pass the same message as a must-do to their children.

Hygienic education is a subject which can be easily incorporated into school syllabi in health education, health science etc. Once it becomes part of a normal routine, it does not require any extra resources. The most effective interventions are those that reach families through children In any aspects of environmental management, moral persuasion has become increasingly popular. It encompasses community management issues, which always require the primary stakeholders are involved in intervention programs even before planning (i.e., at conception stage). This empowers people enough so that they grow with the idea. It eventually becomes a tool that forms part of life, thus automatically guaranteeing ownership and sustainability.

Water treatment is a technology that demands money, extra time and technical know-how. These necessary inputs or resources may not be always available. Secondly, the chemical may not be readily available when needed, and when available, the skill to use it may not be there, and even when both the skill and chemical are there, there may be no money to purchase the chemical. Furthermore, the chemical may impart a taste that may render the water unacceptable. For instance, alum (Aluminium sulphate) is commonly use in water purification. It causes particles to coalesce, coagulate and settle, leading to clarified water. However, Aluminium itself is a health hazard, as it causes human degenerative disorder (Alzheimer's disease) as it is always available in drinking water (Jones, J, 1997). Similarly, the chemical may react with other components of water- e.g. chlorine may react with amines and nitrogenous materials, leading to products and other residues which may be a life hazard. For instance, chlorine forms chloramines and chloramides in water, both of which as carcinogenic. If the water was rich in nitrate, the normal treatment process has no capacity to control it, yet nitrates are known to cause methaemoglobinaemia (blue baby syndrome) in infants of six months and less, who are bottle-fed (Afullo, 1995). In normal water treatment, there is even no provision for eliminating or detoxifying poisonous organic chemicals such as aromatic hydrocarbons, pesticides, heavy metals etc. Thus even in conventional water treatment plants, it is not possible to eliminate dangers associated with water quality Jones, 1997; Afullo, 2003).

Communities are the custodians of natural resources wherever they live and coexist with these resources. They have coexisted since time immemorial, and have always devised means of conserving the resources, minimizing impacts as they go on with their daily activities. They farm, thus cause water pollution by agricultural chemicals; they fish, thus are direct primary stakeholders in river water quality; they hunt, and thus know the dangers they face when the wildlife they rely on for living are killed by water contamination etc. As people with direct stake in water

164

quality, they should be empowered to safeguard against water pollution. Education remains the best way of doing this, so that the fine and complex interactions between water- air- soil-rocks; fertilizer / pesticide- crop- river quality; upstream farming- rainfall-soil erosion-soil impoverishment (low yields)- pond siltation- pond drying; etc can be familiar relationships (Afullo, 2003).

Due to community livelihoods approach to water quality management, it would suffice reinforcing their soil-water-health relationship and water-conservation knowledge base to render the water quality management sustainable. Hygiene education can therefore be tailored to cover a broad base from which he community can easily get the necessary empowerment to manage their most vital resource-water (WEDC, 2003).

Essay 1c: Explain where water quality samples should be taken and why (use diagrams where appropriate)

Importance of water sampling

Water quality samples are normally taken as a part of routine monitoring program to ensure temporal water quality is appropriate for various uses (Afullo, 1995). The best water quality monitoring sites are determined depending on the prevailing circumstances. For in stance, there could be non-point source pollution prevalent in an area. This can be from the sugarcane farms in figure 1. In this case, because of the diverse origin of pollutants (mostly agricultural), it may be difficult to identify the best water sampling point. However, surface runoffs can be sampled from major agricultural water channels, e.g. from major valleys, etc. Alternatively, as part of routine monitoring, we can sample from a major sink such as downstream of a river at the point where the major agricultural practices cease (Afullo, 2003; Jones, 1997 and Afullo, 2004).

Non-point source pollution: In figure 1, this is at point B. In non-point source pollution, there is no definite, identifiable source of pollutants (Jones, 1997). This can give an indication of what has come in from upstream. However, to eliminate background pollutants, it may be worthwhile sampling from two points (points A and B below)- one before and one after the major agricultural zone. Thus point A gives background pollutant levels, while point B gives pollutants directly attributable to the sugarcane farms. This is as indicated in the diagram (Fig 1) below.

Point source pollution: In point source pollution, there is a definite point of discharge of pollutants into a water course (WEDC, 2004; Jones, 1997) (point E in figure 1 below). In this case, if pollution control is the motive, then there is need to sample water at the beginning of pipe (point D in figure 1 below), i.e., before effluent release (i.e., upstream of the industry or effluent source) and at end of pipe, i.e., after effluent release (i.e. downstream of effluent source). This can giver an indication of the impact of the effluent on the receiving water body. It can help determine the dilution rate and efficiency. This is as illustrated in the diagram (Fig 1) below. Point C is a sampling point for monitoring the background pollutant levels before the effluent discharge into a river. Point D is the effluent source at the industry. Sampling and monitoring at source can give an indication of whether or not the effluent meets direct discharge standards, or whether to pre-treat prior to discharge. Since the effluent flows through D to E, there is bound to be some natural effluent cleansing process. Normally, if the channel is long, especially an high temperatures, there is reasonable purification. BOD and microbial purity can be substantially improved. Thereafter, there is need to monitor the effects of the effluent discharge downstream after complete mixing of the discharge with river waters. This sampling point, F, could also be where the public draw their water for use e.g. for domestic or agricultural purposes. Monitoring water quality at point F therefore excludes the possibility of poisoning of man and livestock. Whereas the sampling points shown here is only a guideline. There could be delicate points, e.g. points of ecological concern e.g. swamp with unique and / or sensitive flora and / or fauna which need to be protected Jones, 1997; Afullo, 1995; Afullo, 2003).

Question 2a: Use the information from Jones (1997) and any additional resources you can find to create a clear and concise diagram showing the dominant water inputs, outputs, transfers and storages that are relevant tot the demise of the Aral Sea.

Essay 2b: In no more than 1200 words, describe your diagram and discuss what elements of your diagram can be adjusted to provide possible solutions to the ARAL sea crisis.

(Q 2bi) Key water inputs, outputs, transfers and storages in the Aral Sea basin as shown in the figure above are hereby discussed Inputs: The key Inputs into the Aral Sea are largely rainfall and the rivers Amu Darya and Syr Darya. Underground water also keeps the basal flow. The river flow is maintained by the rainfall and melting ice from the Hindu Kish Mountains and the Tien Shan mountains. The mountains are likely to causing orographic rainfall, as there is large amount of surface water which can be cooled by

166

the mountains and falling within the basin (WEDC, 2004, WEDC, 2004b and Jones, 1997).

Outputs: The key outputs from the sea are: transpiration by crops and other vegetation, evaporation, withdrawal for irrigation, and seepage to underground reservoirs. Indirectly, the addition of salts renders the water less available, as the resulting osmotic pressure associated with the salty water is much higher, necessitating use of more such water for irrigation. This is therefore like a loss, as the resulting basin ends up carrying only solutions and not pure usable water (Jones, 1997; WEDC, 2004).

Storage: The key storages are: underground reservoirs, the Aral Sea itself, and in the vegetation (e.g. trees, crops, grass etc) (Jones, 1997).

Transfers: The key water transfers within the basin are: from surface reservoirs such as Aral Sea, and from the two rivers Amu Darya and Syr Darya to underground; also evaporation to the atmosphere, from where hopefully some of the lost water may find its way into the sea directly or through the two rivers, and in the farms.

Key processes in the cycle:
(i)Evaporation is significant here due to a number of factors which encourage it. These include high temperatures in the desert environment; the relatively strong winds associated with the physical nature of the terrain; large tracts of open land surfaces-all covered by water, etc (Jones, 1997). Together with transpiration, the two are very significant in this basin, as the crops seem to release water freely to the atmosphere without restriction, as there is ready supply to replace the transpired water.

(ii)Evapo-transpiration includes both evaporation and transpiration. The two are significant due to the high atmospheric temperatures, large tracts of open land under crop, and fully flooded with water for many months.

(iii)Storage: The key storage sites are the underground water reservoirs. This occurs through seepage and deep percolation of the surface waters. This in a way makes the water less susceptible to evaporative losses, surface pollution etc. However, the groundwater is more prone to nitrate pollution, as the area is rich agricultural zone where heavy nitrate-fertilizer use is promoted and practiced. This exposes the groundwater to pollution by the highly water-soluble nitrate fertilizers often applied in rice fields (Jones, 1997; Afullo, 1995).

The underground seepage, however, renders the surface water less available, and has greatly contributed to the certain death of the Aral Sea. The seepage is worsened by the reduced water flow, reduced river discharge (from combined water withdrawal fro irrigation, and evaporation), increased erosion and associated elevated levels of suspended solids (TSS). The latter are largely fine silt, clay and sand from agricultural fields. These get deposited along the river banks, ending up reducing the river channel. The case illustrates how the river Syr Darya was partly blocked by the silt. Naturally, as siltation continues amidst reduced river discharge, the river may get blocked as the suspended load increases, and evaporation also increases viscosity of water in the river, and thereby significantly reducing its velocity. This leads to increased residence time of water at specific sites, which in turn increase evaporation and seepage through the sandy soil beds of the Aral Sea basin. Thus the water gets stored underground at the expense of the surface water. Some of the water also seeps from the large tracts of agricultural farms where long residence time encourages seepage and evaporation (Jones, 1997; Afullo, 2004; Afullo, 1995).

(Q 2biii) Possible interventions to solve the Aral Sea crisis
Introduction: Already a number of interventions are proposed by Jones (1997). As the author rightly observes, the interventions need to be of mega scale, as the problem itself is a life-threatening one. No amount of investment is too large if it is meant to save lives. Thus any of the interventions suggested by Jones can be tried. However, I will suggest some, perhaps even similar to what Jones suggests. The alternatives are discussed below.

(a)Desilt Rivers Amu Darya and Syr Darya. This is likely to increase flow to the Aral Sea. This process should, however, be accompanied by soil and water conservation measures within the rice and cotton plantations which actually contribute the solids. In essence, as the siltation is halted, the impoverished soils will also get rejuvenated. Methods that can be used to conserve the soils of the basin include: gabions, terraces, waterways, dams / reservoirs etc. Gabions can be used to gradually eliminate gullies within the basin; the dams can be used to stabilize the silt-ladden waters so that the suspended matter settles and the remaining clarified water can be used for irrigation.

(b)Lagoons can be constructed to keep irrigation wastewater / effluents so that it gets naturally treated. Some aquatic plants (floaters) can be introduced into the lagoons to reduce the loads of some of the pollutants before the water is recycled. Lagoon design can be done with a minimum residence time in mind, just like the waste stabilization ponds normally are in effluent treatment (WEDC, 2003).

© Crop rotation can be practices to give the impoverished soils time to recover, as well as reduce the intensity of farming on the land. For instance, paddy rice can be rotated with a crop which does not require flooding. This can reduce the demands on the available water, while new crop utilizes the agrochemical residues in the soil to produce food for the millions of residents. This can also help reduce over-reliance on only one crop and food (i.e. monoculture). With monoculture stopped, there is bound to be a change in pest and weed scenario. A change of pesticide and / or herbicide can help reduce water pollution. Some of the new crops being introduced can even be selected based on their resistance to drought, pests and tolerance to weeds (Afullo, 1995).

(d) Change of farming technology and reduce intensity of farming in the basin

Manual methods and other alternative weed and pest management methods can be introduced, in the spirit of integrated pest and weed management. This can not only reduce the intensity of land use, but also rejuvenate it. It would be better if non-mechanized farming methods were resorted to, perhaps in combination with mechanized methods. Thus can help reduce the physical manipulation of soils by the heavy machines (e.g. tractors, rotovators, combine harvesters etc)- which often destroy the soil structure- and render it more prone to erosion, flooding, and loss of soil-stabilizing cations such as calcium and magnesium. Once the physical strain on the soil is reduced, the cations can be more balanced, and the soils can recover faster, losing less, if any silt, and making the surface waters safer (Afullo, 1995).

(e) Change of irrigation technology: Flood irrigation as used in paddy farming is very water inefficient method. Alternative methods such as drip weep and trickle irrigation methods can be used- on other non- hydrophilic crops. This thus calls for change of crops, as well as change of technology to help reduce the strain on available water resources. More intensive and water-efficient farming can be done by change of or rotation of crops. The new technology can still even work on non-paddy rice (i.e. highland rice which normally neither requires less water, and therefore does nor require flooding). This can be preceded by crop breeding research and trials so that only stress-tolerant (pest, heat, drought and weed-tolerant rice / crop) variety / strain replaces the current one (Jones, 1997). This can reduce chemical and fertilizer loads to the water bodies, besides conserving the little available. This can give the lagoons time to purify the effluents for use in the new crop or rice variety (Afullo, 1995; Afullo, 2004).

(f) Integrated water management so that the polluted sources can be conserved and rejuvenated, while clean sources are used to serve the public. This is where the withdrawal from Caspian Sea and / or Volga or Siberian river would apply. Integrated water resources management (IWRM) emphasizes the need to collaborate closely among neighbouring basins to ensure water quality and quantity are conserved (Jones, 1997). This also means having understanding on when and how to share resources when need be, including importing from another basin if necessary. This situation of Aral basin crisis requires inter-basin management approach, as clean water has to be acquired from elsewhere even for a short time while the polluted sources are getting rejuvenated. This may look expensive, but it is cheaper that the many water pollution -related diseases that human beings are facing in the Aral basin.

(g) Forestation: Trees can be planted as part of water and soil conservation, but given special attention since it has capacity to solve a lot of advantages in balancing the ecosystem and stabilizing water resources.

Conclusion: The solution to the Aral Sea crisis lies in IWRM approach (Jones, 1997; WEDC, 2004), as the water quality and quantity need some time to repair using longer term interventions. A combination of soil, water and other technological approaches are necessary to solve the elaborate problem, which needs not be given a quick-fix solution. Simple incremental yet inevitably expensive approach to the problems can be helpful.

Essay 3A: Discuss in no more than 300 words the reasons for considering water resources at a municipal level based on the boundaries of the urban area rather than boundaries of natural catchments.

Introduction

A catchment is a geographical area surrounded by recognisable boundaries within which the flow directions of water converge to a common exit (Jones (1997). WEDC (2004) suggest catchments as the ideal unit for managing water resources. Indeed, this is true, from the technical and realistic point of view. Agenda 21 of UN conference on environment and development also says that 'IWRM, including the integration of land and water-based aspects, should be carried out at the level of the catchment basin or sub-basin'. Catchment approach to water resources management therefore seems to be widely accepted. Yet this is based majorly on theoretical grounds.

The reality: IWRM is multi-facet

In reality, there is no one way of managing water resources. Logistics and practical perspectives make this catchment ideal unworkable (Jones, 1997). For instance physical features such as rivers, escarpments etc often form the boundaries of most administrative units. Administrative bureaucracies cannot enable efficient management of resources, as management vary from one to the other- some with a foresight and plans, and others managing otherwise. Managing such resources need close collaboration among administrative units. However, a number of neighbouring tribes, countries and even clans are known not to be in talking terms. This can complicate matters if life-saving and livelihood resources such as water are left at the mercy of such administrative units. Different administrative units also have different priorities, plans and programs which are more or less dictated by their provincial, regional or national priorities. It can be very difficult to reconcile these with the priorities of a neighbouring administrative unit if they are in a different administrative zone- even if they share a catchment (Jones, 1997). These are therefore best managed independently from a practical perspective.

Water resources and international environmental calamities know no bounds

Secondly, a number of international or regional environmental crises such as desertification, pollution etc is not necessarily of catchment in nature (WEDC, 2004). Yet they affect climate over a wider area. Restricting management to a catchment therefore cannot be feasible in such situations. Similarly, different uses are put to different water sources. For instance, dams, taps, rivers, lakes, swamps, etc are used for different purposes. Due to diverse interests in different water resources and their users, it is not possible to manage as a catchment. Underground water resources may not even be from the same geographical area where it is being used; and the resource may not even follow the surface water catchment pattern (Jones, 1997; WEDC, 2004). The base flow which partly stabilizes the surface water resources in a catchment may originate from an exogenous underground water resource. Thus conserving such water requires that other catchments be considered. Since management is for purposes of user, professional, political, strategic and related interests, diverse interests can make management base on a unique geographical unit very difficult.

Conclusion

Therefore before deciding on a particular management unit for water resources, it is better to know the facts of water in a given area- including geological, hydrological, hydro geological, political, international, and related perspectives to water management. The catchment approach may look most reasonably the best on face value, but a thorough assessment may prove otherwise in some situations. In a municipality, it is better managed at municipal boundary level.

3B: A simple diagram of water cycle based on an urban area. Show flows of water in and out of the area and major stores of water.

HYDROGRAPH

Essay: Discuss in no more than 300 words how the shape of a hydrograph is influenced by physical (natural and man-made) catchment characteristics, such as area, slope, shape and drainage density. Use clear and appropriate diagrams to support your discussion.

A hydrograph is a diagram in an x-y plain in which stream flow variation is recorded over time, i.e., stream flow is plotted against time in a given area. This can be useful, and has been used to determine and predict similar occurrences, e.g. floods. The area under the hydrograph minus the base flow represents the volume of runoff. Since it is runoff which turns into floods, a hydrograph can be used to predict the possibility of a flood.

Essay: How the shape of a hydrograph is influenced by physical (natural and man-made) catchment characteristics, such as area, slope, shape and drainage density.

Slope
Slope of a catchment influences the shape of a hydrograph by the speed of water flow (Jones, 1997). The higher the slope, the faster the water flows, thus giving a sharp peak of a hydrograph. The flow is not extended over a long time, bur only occurs during the rain. The flow stops almost immediately the rain stops, while it also takes a short time from start of rainfall to start flowing. Sloppy catchments thus give very long but squeezed hydrographs. See figure 5 below..

Drainage density
High drainage density implies there are more channels of flow (Jones, 1997). The higher the density, the more the chances the different flows emerge fast, and increases volume of flow. This may therefore soon reach a peak, which can extend for a long time even after the rain subsides. High density catchments thus give long and wide hydrograph. See figure 6 below.

Catchment Area
A large catchment is likely to have many micro-catchments, which means more influence from internal and neighbouring features is highly likely. Thus the height of the hydrograph may vary from one micro-catchment to the other, giving the overall maximum intensity that varies widely from low to very high. The time base of the flow is also likely to be rather long (Fig 7).

Essay: briefly discuss why it is important to be aware of surface flow rates of water, giving examples of surface water resources with very different flow rates.

Introduction: The surface flow includes surface / overland flow (or runoff) and river/stream/channel flows.

Overland flows: The overland flow measurements can be useful in predicting the flood hazard in an area. It can also be useful in determining the possible water reserves in an area if this water was to be stored in a dam or any other reservoir. The surface flow also indicates the nature of soil and rocks in the area. High overland flow may mean either saturates soils, or low water

173

permeability of the soils. The latter is possible if the rock strata are impervious, indicating a possibility of flooding unless the excess overland flow gets a sink. Lastly, the overland flow itself can be an indication of the amount of precipitation in an area. An area with high precipitation is most likely to also have a high overflow (Jones, 1997; WEDC, 2004).

River / stream/ channel flow: On the other hand, channel, stream or river flow indicated a number of characteristics of the catchment. Perennial flows mean there is continuous supply of water to the surface water resource, whose base ids supported by base flow. Perennial flow also indicates the place is neither arid nor semi-arid-places where normally the rate of evaporation far exceeds the base flow, which therefore never exists. Seasonal (ephemeral and short-term) rivers may depict a type of underground rock (perhaps too pervious) and upstream climate. All these factors together are useful in predicting natural calamities such as landslides, floods and even droughts. If the records are kept over time, they can give a time series- a pattern which can be useful in predicting major flow changes. It may be possible to plan to avert crises, for instance if an area is known to flood after every 10 years, the 10 years after one flood can be used to plan and implement a major intervention (Jones, 1997; WEDC, 2004)

REFERENCES
1. Afullo A (2004) Environmental and occupational health aspects of waste management in Maun, Okavango Delta, Botswana. An Unpublished PhD Thesis. COU, Spain.
2. Afullo, A. (2003) The environment- some concepts, issues and concerns. Sustainable Futures Maun, Botswana.
3. Afullo A (1995). Pollution of Lake Victoria by inorganic fertilizers used in the West kano rice irrigation scheme, Kisumu, Kenya. An Unpublished MPhil Thesis, Moi University, Eldoret.
4. Durham N and Jackson M (2001) People and systems for water, sanitation and health: Monitoring water quality in the developing world. 27th WEDC conference, Lusaka, Zambia
5. Ince M and Howard G (1999) Integrated development for water supply and sanitation: Developing realistic drinking water quality standards. 25th WEDC conference, Addis Ababa, Ethiopia
6. Jones J A (1997). Global Hydrology: Processes, resources and environmental management. Pearson prentice Hall, ISBN 0 582 09861 0. Essex.
7. Von Sperling (1998)
8. WEDC (2004) Integrated Water Resources Management (IWRM): A WEDC Postgraduate module. WEDC, Loughborough University, UK.
9. WEDC (2004b) Integrated Water Resources Management Topic task 2. WEDC, Loughborough University, UK.
10. WEDC (2003) Community and management: A WEDC Postgraduate module. WEDC, Loughborough Univ, UK

11. WEDC (2003b) Wastewater Treatment: A WEDC Postgraduate module. WEDC, Loughborough University, UK.

Topic task questions and comments on your answers	grade awarded
Question 1	
a) in no more than 2 pages, discuss the importance of social, human, financial and environmental factors in managing water quality. You do not need to include technical and scientific issues, unless they relate to one of the four factors listed. Some comments about the text The society – Societal Traditionally there were – maybe this statement and the following text needs a specific context? Perhaps a case study example, a region, or a specific society? Some sides of water monitoring points – could easily be made more specific Always water of the right quality – sounds good, but in which context should this be placed and what is the lesson learned from this? Putting your following sentence in this context – is only a homogeneous society successful in managing water quality? What about gender, social status, water quality acceptability (smell/colour), quantity versus quality, expectations, etc.? Environment Components poisonous – is components here a component of the water cycle or a substance in a water body? Rives – rivers This section could do with tightening up a bit. Providing a better structure (perhaps even a table with a list of variables to be considered and followed up by a description of some of the major elements of the table could be more useful). Other environmental aspects to be considered would include catchment conservation, climate (environmental) change, environmental legislation, sustainable development, etc. Sloppier? – environments with steeper slopes, higher relative relief, more undulating terrain? Human Ok a fair range of issues has been raised, but there are also issues to be considered such as the right to safe drinking water, affordability, and end-user involvement. Perhaps by using some more generic subheading in this section, you could have grouped some of your discussion topics, while at the same time allowing a broader approach to the topic. Financial In addition to construction, O&M and credit facilities other aspects such as affordability (again), financial sustainability, prevention cheaper than cure, water pricing, etc could be discussed in this section. The references could have been somewhat better. In a discussion like this it is the use of additional resources that makes it more interesting to analyse. Although the question specifies the discussion of four factors, it would have been useful to conclude in your answer that an approach integrating all four (and other) factors should be followed.	
b) In no more than 2 pages, discuss "Water quality is best improved by increased hygiene education rather than treatment" Good answer, spirited text, well written and convincing. Well done.	A
c) In no more than 1 page, plus diagrams if necessary, explain where water quality samples should be taken and why. Good range of examples, and well explained. There are, naturally, many more different scenarios possible where water quality sampling is appropriate, but we're aware that you cannot possibly discuss them all in one page.	B+

The where and why are both considered well.	

Question 2

i) Use the information from Jones (1997) and any additional resources you can find to create a clear and concise diagram showing the dominant water inputs, outputs, transfers and storages that are relevant to the demise of the Aral Sea.	B

ii)   In no more than 1,200 words describe your diagram and discuss what elements of your diagram can be adjusted to provide possible solutions to the Aral Sea crisis.

Using a diagram, rather than a map, you have a number of opportunities to highlight certain issues in this region. If you just want to mention the distribution of inputs, outputs, transfers and stores, it may still be nest to use a close representation of the geography of the region as a basis. In this case, the diagram could be improved by showing more clearly the geographical layout of the Aral Sea region. If it was intended, however, to show the relative importance of the various components in the diagram (a preferred approach), some sort of scale factor should be added. For example, your area of evapotranspiration losses in the irrigation 'bubble' is quite large, suggesting that this is relatively more important than the losses to groundwater. Although many components are shown the diagram is not very informative. It does not clearly show how Syr Darya and Amu Darya contribute to Aral Sea. The diagram misses a number of important issues, including a representation of the freshwater surface flows that originate largely in the Kyrgyz mountains (inputs ice melt, snow, rain, storage snow and ice, transfers rivers, glaciers), a representation of the relative amounts of water volume transfers via rivers, river diversions, canal networks, irrigation channels to agricultural basins where outputs (evapotranspiration) and storages (groundwater) exist and finally further transfers via surface channels leading to inflow into Aral Sea (with its own balance of inputs, storages and outputs). In the subsequent discussion of the diagram and the issues involved, there is no hard data mentioned that underpins the diagram and the problems faced in this region. The internal drainage of this region could have been more clearly shown in the diagram.

The discussion is reasonably well structured but could do with tightening up a bit. You could have combined the explanation of the diagram with the discussion about the problem and possible solutions. Also try to reduce waffle (in the introduction for example where you state that any of Jones's solutions could be tried, and that you will suggest some, perhaps even similar interventions..... This is unnecessary.

It is interesting that IWRM is mentioned as a separate option. Wouldn't it be better to fit all solutions under an integrated umbrella? The discussion could then become much more concise and targeted. I miss 'hard' data illustrating the demise of the Aral sea (a comparison of water budgets from 1960 and 1990 would have helped here) and assessments of how much each fo the solutions could contribute to the reduction of the current problems. As a consequence the reader does not get much of an idea what the scale of the problems and the magnitude of the required solutions are.

Question 3

a) discuss in no more than 300 words the reasons for considering water resources at a municipal level based on boundaries of the urban area rather than the boundaries of the natural catchment(s)	B+

This is a good answer. I like the comment that there is no such thing as one good way and that IWRM is multi-faceted. However, I find it difficult to see that the urban scale is the answer to the comment that large scale environmental crises are not necessarily approachable from a catchment scale, that does not seem to fit... Management options at these scales should rather take the greater context into consideration.... Perhaps it would be easier to have discussed the remainder under the headings of politics, finances and organisational structure.

Some advantages of an urban (municipal) approach include (not in order of importance and some you touched upon already): • More appropriate water allocation • Better dispute management with neighbouring administrative areas over water rights and water use • More effective regulation and policy development • More appropriate water pricing • Greater accountability • Better monitoring • Greater understanding of local network and functioning of water management infrastructure	
b) draw a simple diagram of a water cycle based on an urban area. Show flows of water in and out of the area and major stores of water. I find this diagram difficult to interpret. It's full of unexplained arrows, a mix of detail and generalisation, an unclear set of pathways and unnecessary detail. The use of a water budget to explain a diagram like this good be a good starting point. There are of course many different ways in which you can draw an urban water cycle. One approach to get more information in a diagram and convey this information clearly to the reader, you could have a look at the Lvovich (1977) diagrams as represented in Jones (pages 16 and 17).	D+
c) Discuss in no more than 300 words how the shape of a hydrograph is influenced by physical (natural and man-made) catchment characteristics, such as area, slope, shape and drainage density. Use clear and appropriate diagrams to support your discussion Figure 4 is pretty bad. There is no info on the y-axis and the explanation only helps a little. Much more detail could have been given (see your DL document). Simplifications such as runoff turns into floods is not especially helpful in this context. The three following diagrams are taken from Jones and are relevant in this discussion, the one in Jones is a bit better, though......... Try to expand on Jones's stuff by adding info from other literature.	C
d) Briefly discuss why it is important to be aware of surface flow rates of water, giving examples of surface water resources with very different flow rates introduction is a bit short...... It is more usual to set out what you're about to discuss, rather than a statement of a definition.... You're getting into the right direction, but it could be improved by firstly defining flow rates as volume over time (e.g. m^3/s). Separate this in the discussion from flow velocity (m/s). These are not the same. Greater flow rates result in greater water renewal rates and are important for, for example, determining sustainable water consumption rates or analysing the nature of natural of man-made catchments. Water consumption of surface flows must be in balance to be sustainable and very slow renewal times may rapidly put extraction rates beyond the capacity of the resource and ultimately resource degradation will occur. Environmental resilience is related to renewal. Other issues such as environmental quality of catchments can be gleaned from surface flow rates. If flow rates rapidly respond to changes in inputs (e.g. rainfall) this could indicate that the natural catchment has limited infiltration and interception and if the shape of the catchment induces high flow velocities the erosion capacity of the flow may be significant.	C

IWRM TOPIC TASK 3
CONJUNCTIVE USE OF WATER
Essay: Conjunctive use of water resources: a case study of Kasau Village, Lake Victoria basin, Kenya

Executive summary

Conjunctive water use is the integrated use of diversity of water resources with a view to increasing combined benefits or achieving a more equitable distribution of benefits from these resources (WEDC, 2004). Availability of water of various sources to a community depends on topography, hydrology, physiognomy, culture, and economic activities demanding water. This paper presents a case study of Kasau village, Bondo District, Lake Victoria basin, Kenya. This lies in the lowlands of western part of Kenya. The main water resources in Kasau village are Ponds, river Yala, Lake Victoria, the wetlands, borehole, ephemeral water in pits, and roof catchment.

Roof catchment is largely spared for drinking, and is available during rainy season. Its collection is done using the eaves of grass-houses, and corrugated iron sheet-roofs. The major ponds are Okelo, Sitina, Tinga and Afulo. Tinga is permanent and all-purpose; Okelo is for drinking; while Afulo and Sitina ponds are seasonal and all-purpose water sources. There is one borehole (Naikonda) constructed in 1988 by the Lake Basin development authority (LBDA). Its salty water is not popular with the community. It is therefore only used during dry season when ponds are dry.

The village is littered with pits formed by mining of gravel for building. These occasionally fill with rain water, and are commonly used by the homesteads for 2-3 weeks during the rainy season. They are found within homesteads and therefore very popular during rainy season. They are never used for drinking. The Lakes Kanyaboli, Sare and Nyamboyo are the wetlands. They are permanent, muddy, and comprise floating plants such as papyrus. They are neither accessible to the majority nor readily accessed by carts, bicycles and wheelbarrows, rendering them only for small scale use by fishermen and herdsmen. Nearby homesteads use them for bathing and fishing.

Lastly, there is the permanent fresh water resource- Lake Victoria. Its clear water is used for drinking, among others, throughout the year. However, it faces dangers of pollution and its quality is currently low. It also faces the water hyacinth (Eichhornia crassipes) crisis which tends to make it shallower. It is a permanent supply with relatively clarified water. Handcarts, Bicycles, wheelbarrows and use of head are the means used to deliver the water to homesteads, either by members themselves or by private water vendors. This limits its use especially by poor families, who then rely on family labour to bring the water. As a coping strategy, they bath and wash clothes directly in the lake waters. Water is only brought for use in cooking, drinking, washing of utensils, and bathing of infants, elderly adults and some men.

At any given time, more than one source of water is used by each homestead, depending on the accessibility and perceived cleanness of the source. These sources vary with season, which are mostly defined by rains and drought.

Introduction
Conjunctive water use is the integrated use of a diversity of water resources with a view to increasing combined benefits or achieving a more equitable distribution of benefits from these resources (WEDC, 2004). These sources could include surface (river, lake, dam, and pond), underground (wells, boreholes), springs and roof catchment. A mix of use of these resources depending on their spatial and temporal distribution to a population improves the general welfare, in terms of time saved, minimization of competition among various demands on specific resources, reduces price per unit of water (as competition for various sources / quality is reduced), etc. In this way, water can be used as a means of reducing poverty through more rational; use for various economic activities such as agriculture, industry, etc. Availability of water of various sources to a community depends on many factors, the major ones of which are: topography, hydrology, physiognomy, culture, and economic activities demanding water (Jones, 1997 and WEDC, 2004). A case for study here is the Kasau village, Bondo District, Lake Victoria basin, Kenya. This lies in the lowlands of western part of the country, within the Lake Victoria basin.

The Lake Victoria drainage basin
This covers an area of 266,000 km^2, covering Kenya, Uganda and Tanzania. Of this, Lake Victoria constitutes about 68,000 km^2, Islands cover 3000 km^2 and mainland 195,000 km^2 (GOK,

1994; Otieno and Okidi, 1992). The Kenyan side of Lake Basin is the area of study. This covers a land area of 4,000 km^2 (Afullo, 1995; GOK, 1994; Otieno and Okidi, 1992). This lies to the western parts of the deep slopes, the Mau range down to Lake Victoria. It covers Nyanza, Western and parts of Rift Valley provinces.

Climate of the Lake Victoria Drainage Basin VS water availability
The Lake Victoria drainage basin has a mean annual runoff of 7292 million m^3 and an average rainfall of 1245 - 1342 mmpa (GOK, 1994; Afullo, 1995), and a mean annual evapo-transpiration of 1096 mmpa. The climate varies widely with altitude. The low lands near the lake shores receive low, unreliable rain, with the driest zones receiving 500 - 1000 mmpa. This also covers Kasau village, the study area. Rainfall is bimodal, with peaks falling in May and October. The short rain largely coming later in the year is unreliable. The temperatures are high, generally averaging 22 - 24 ^{0}C. In some cases, even 30 ^{0}C has been recorded.

Generally, the temperatures change with time of the year. The sunshine hours in these areas are averagely 8.3 daily. There are more sunlight hours as one nears the lake, and fewer, further away. The relative humidity is 85%, 68% and 47% at 0300, 0600 and 1200 hours respectively. In Kasau village, like most other lower altitude areas of the lake basin, the evapo-transpiration almost always exceeds precipitation, giving net negative moisture availability (Obara and Ogonda, 1990). Regular crop failures are therefore a common scene.

WATER SOURCES: GENERAL:
The main water sources in Kasau village are: Pond water, the river Yala, Lake Victoria, the wetlands, a borehole, ephemeral water in pits, and roof catchment. Roof catchment is done using the eaves of grass-houses, and corrugated iron sheets. This largely used for drinking. Kasau primary is a community school whose facilities (houses and roof catchment tanks) have all been built by the community. It has a well-designed roof catchment system, with two large tanks. Teachers and pupils use this water in the school. However, the community uses the water for all home activities- with permission from the head teacher- during funerals and festivities within the village, especially when the tank is full. During dry season, only a little can be given to any one household in need (i.e., there is

rationing). This is to spread its availability until the next rainy season.

Ponds:
The major ponds are Okelo, Sitina, Tinga and Afulo. Tinga is permanent and all-purpose (except drinking). Okelo is for drinking and other less-water demanding household activities. Afulo and Sitina are seasonal and all-purpose, except drinking. There is one borehole (Naikonda), which was constructed in 1988 by the Lake Basin development authority (LBDA). Its salty water is not popular with the community. It is therefore only used during dry season when ponds are dry (personal observation).

Gravel Mine spoils and the wetlands
The village is littered with pits formed by mining of gravel for building. The gravel pits present best example of subsidiarity (WEDC, 2003; and WEDC, 2004). The pits often fill with water, and are commonly used by the homesteads during the rainy season, as they are seasonal. The advantage is that they are found right within homesteads, near households, making them the water sources of choice during rainy season. They are, however, very short-lived, and often dry up within two to three weeks after any heavy rain. They are never used for drinking, but serve all other water-demanding purposes. The wetlands include Kanyaboli, Sare and Nyamboyo Lakes. They are permanent, but muddy, and comprising floating plants such as papyrus. They are therefore not accessible to the majority. Nearby communities use it for bathing and fishing. Other uses come in during dry season. They cannot be readily accessed by carts, bicycles and wheelbarrows, rendering them only for small scale use by fishermen and herdsmen. The table below shows the preferred and actual use and seasonal distribution of these water sources.

LAKE VICTORIA
Lake Victoria covers an area of 69,000 km^2 of which about 4,000 km^2 is in Kenya (GOK (1989a; GOK (1989b; GOK (1993a and GOK (1993b). It is therefore the second largest fresh water lake globally (Obara and Ogonda, 1990). It has a maximum depth of 60-90 m (Average 82m) and an average depth of 40m. The Winam Gulf, which forms a good proportion of the Kenya side, is only averagely 7 m deep. This puts it at great risk of being prone to eutrophication consequences: early death. Due to

pollution from agriculture, industries and municipals within the basin, the lake water has been seriously compromised (Nyamu, 1986; Otieno and Okidi, 1992; and Mbuthia, 1987).

As a result, it has faced the water hyacinth (Eichhornia crassipes) crisis for the last eight years (Afullo , 1995). This has improved the chemical water quality a little bit. Households that can reach it generally use the lake water universally for washing, bathing, drawing for domestic purposes, livestock drinking (direct), etc. Those who live far from it use it during dry seasons when alternative sources (ponds) are dry. Its main advantage is permanent supply and relatively clarified water. It is therefore only filtered and / or boiled at homesteads for drinking. Handcarts (with drums, Jeri cans), Bicycles (with Jeri cans), wheelbarrows (with Jeri cans) and use of head (Jeri cans and pails / buckets) are the means used to deliver the water to homesteads, either by members themselves or by private water suppliers who sell it.   This limits its use especially by poor families, who then rely on family labour to bring the water. As a coping strategy, they bath and wash clothes directly in the lake waters, and then only carry them home to dry. Water is only brought for use in cooking, drinking, washing of utensils, and bathing of infants, elderly adults and some men.

Other factors that are relevant in the Lake Victoria basin that endanger the Lake Victoria water quality and lifespan:-
♦   Uncontrolled siltation arising from land abuse;
♦   Eutrophication arising from agrochemical abuse;
♦   Heavy metal toxicity arising from irresponsible disposal of raw industrial wastes (Alala, 1981);
♦   Low microbiological (Bacterial) quality arising from wanton disposal of raw sewage (Mbuthia, 1987);
♦   Unbalanced ecological composition arising from pesticide abuse (Nyamu, 1986);
♦   Reduced water levels rising from unsustainable luxurious withdrawals;
♦   Soil impoverishment following soil erosion, inappropriate farm husbandry and deforestation (Afullo, 1995).

Already these have caused clear signs of ecological disaster (Afullo, 1995 and Otieno and Okidi, 1992). Such as:-
♦   Invasion of the water hyacinth (Eichhornia crassipes) which since 1996 has been a problem;
♦   Serious soil degradation, leading to badlands, low and

negligible crop yields, persistent crop failures and general pathetic poverty in the basins
♦ Regular flooding at the lower zones of rivers such as Yala swamp and Kano plains, attributed to siltation from upstream
♦ Greatly narrowed fish diversity from over 100 species to 3 species;
♦ Blocking of fishing points / beaches by water hyacinth. (NB The same fishing boats carried jericans for drawing clean drinking water for household use)

THE LAKE VICTORIA WETLANDS

River Yala forms the Yala swamp at its delta. This forms the wetlands (smaller lakes) such as Kanyaboli, Sare and Nyamboyo. Kanyaboli is an ox-bow lake formed from river Yala. These are rich sources of fish species that have left Lake Victoria following major ecological disasters (Alala, 1981; Otieno and Okidi, 1992 and Afullo 1995). They are wetlands also heavily covered with floating papyrus plants (Cyperus papyrus). These render these wetlands less useful as potable water sources. The neighbouring households use them in fishing, occasional bathing, and livestock drinking.

River Yala:

This is a permanent water source. GOK (1993a) and GOK (1993b) recognise this as a very useful resource for the area's development. However, it is only available to those living in proximity to it. These people use it throughout, for almost all uses. Those a bit far may temporarily reduce using it during the rains, and resort to ephemeral standing waters in pits and depressions. The river water, however, is heavily polluted by industrial chemicals, from the industries upstream (Webuye Paper mills and the sugar industries) (Afullo, 1995; Alala, 1981; Jensen, 1977 and Otieno and Okidi, 1992).

Siltation from the upstream farms has also made the water very brown; some pesticides used in the upstream plantations also pollute the water; while the bacteriological quality is also questionable, as there is rampant defecation in the bushes as well as at the washing and bathing points adjacent to the river (GOK, 1993a, GOK, 1993b, GOK, 1989a and GOK, 1989b). So the use of the river is largely because of its permanent supply, and proximity to many households. Water is drunk directly, or clarified using Alum. Bacterial quality of the water is assured by boiling.

The Ponds (Okelo, Sitina and Afulo):
These are well distributed in the village (Kasau village), and other villages also have as good distribution of other ponds. Ponds are the most popular but seasonal water sources, used as long as they have water. Only Tinga pond (discussed later) is permanent. Okelo, Afulo, and Sitina are seasonal (Even though they used to have water for about 10 months in a year in the 1970s and 1980s, but have been made shallow by siltation). They fill during rainy season, and dry up for a month or two before the next rains come. They thus supply water for general use for at least 9 months in the year. They are largely used for bathing, washing and livestock drinking (except Okelo which is preserved for household use).

The water is turbid, and generally has to be treated with Alum at the household before use in cooking and washing of clothes. It is never used for drinking. Thus almost throughout, their use is supplemented with an alternative drinking water source (borehole, Okelo pond, Lake Victoria, or River Yala). They also dry up faster due to the tragedy of the commons syndrome- as they are used by almost all neighboring villages for livestock drinking- which worsen soil erosion and siltation (personal knowledge of the area).

The Okelo Pond (was closely community managed in the 1980s; this has relaxed in the last 5 years).
Okelo pond is (and has been, since mid 1980s) specifically for household use, and strict control of use (was) is in place. The community employed a member (in the 1980s and early 1990s) to guard it, and ensure its gates were closed in the evening (7-8 pm), and opened in the morning (6 am). The custodian also controlled use, entry into the water, desilting activities (work-for water) etc. This made it less liable to siltation. Its water therefore remained relatively clear, and the pond stayed longer than the other seasonal ponds deep into the dry season. The custodian died, and this led to a vacuum, which has not been filled to date. The community is therefore more prone to drinking water scarcity, as they have to get it from Lake Victoria, or River Yala (personal knowledge of the area).

The Tinga Pond;

This is currently the only permanent pond in the whole village. It is used for livestock drinking, but also at time for bathing, building. Since its water is very turbid, It is only used for cooking and washing after treatment with Alum (dawa), which clarifies it. Some families also use it for drinking after clarification. The tragedy of the commons, and heavy deforestation, however, is degrading it very fast (GOK 1994a; and GOK, 1994b). It is heavily silted, and at times, cattle get stuck in the mud. It is used by at least two villages as cattle drinking point, a process that has resulted in gully erosion and siltation (personal knowledge of the area). It is therefore unlikely to be useful in the future unless it is desilted. The dangers it faces are likely to put the entire community in jeopardy, as they will have to resort to traveling longer distances to take their livestock for drinking.

Conclusion
Generally, Kasau village has a good distribution of water resources during rainy season, quality notwithstanding. The different sources are used depending on their accessibility, quality and convenience. At any given time, at least two sources are used for different purposes. The existing current siltation of ponds threatens to reduce the number of alternatives.

YYYY- Spared specifically for potable use; desiltation done by community during dry season; and restriction in ways of drawing water (no stepping inside); manned by community own hired person (a community member).
YYY- The most commonly used / popular source;
Y- Used simply because of proximity of household to water source, but is not the preferred choice for the specific use (may be because laborious e.g. drawing borehole water for livestock).
Y** Water used at source; water drawn and used near source;
Y*** Water used directly from source- livestock taken to drink directly from source, or clothes washed in the river directly;
y- Sometimes used, especially by children
Ya- used after some treatment e.g. application of alum (locally known as dawa); or boiling;
XXX means use depends on proximity of household to the source; some use it throughout, while others only use it in dry seasons;
Intensity of colour means intensity of source use; grey shading means used by a few households, or used for only selected uses and an alternative source has to be sought to fill the gap; Black shading means universal use by households; or total reliance of the

source by the households for all water-demanding livelihood activities.

References
1. Afullo A (1995). Pollution of Lake Victoria by inorganic fertilizers used in the West Kano rice irrigation scheme, Kisumu, Kenya. An Unpublished MPhil Thesis, Moi University, Eldoret, Kenya.
2. Alala L. (1981). Heavy metals concentration in Kenyan Lakes. An Unpublished MSc Thesis, University Of Nairobi.
3. GOK (1989a) Siaya District Development Plan. MPND and Government Printers, Nairobi. Pp 4-83.
4. GOK (1989b) National Development Plan. MPND and Government Printers, Nairobi. Pp 11-174.
5. GOK (1994a) Kenya Forestry Master Plan Development Program. MENR and Government Printers, Nairobi. Pp 1-420.
6. GOK (1994b) Kenya National Environment Action Plan report. MENR and Government Printers, Nairobi. Pp 1-168.
7. GOK (1993a) Siaya District Development Plan. MPND and Government Printers, Nairobi. Pp 1-84.
8. GOK (1993b) National Development Plan. MPND and Government Printers, Nairobi. Pp 1-124.
9. Ince M and Howard G (1999) Integrated development for water supply and sanitation: Developing realistic drinking water quality standards. 25th WEDC conference, Addis Ababa, Ethiopia
10. Jensen A (1977). Analysis of Lake Victoria Water, Bass fish chlorinated hydrocarbons. University Of Nairobi. Unpublished MSc Thesis.
11. Jones J A (1997). Global Hydrology: Processes, resources and environmental management. Pearson prentice Hall, ISBN 0 582 09861 0. Essex.
12. Mbuthia B (1987). A study of tannery effluents and their effects on receiving waters. Unpublished MSc Thesis, University
13. Nyamu A (1986). Assessment of Water quality in the Kisumu bay of Winam Gulf, Lake Victoria, Kenya. An unpublished MSc Thesis, University of Nairobi.
14. Obara D and Ogonda R. (eds) (1990). Kenya Secondary School Atlas. Macmillan Kenya Ltd., Nairobi. Pp 1-25.
15. Otieno D and Okidi C (1992). Water pollution in Lake Victoria: Causes, effects and abatement. An overview paper, Kisumu. LBDA and FAO.
16. WEDC (2004) Integrated Water Resources Management (IWRM): A WEDC Postgraduate module. WEDC, Loughborough University, UK.
17. WEDC (2003) Community and management: A WEDC Postgraduate module. WEDC, Loughborough University, UK

OTHER QUESTIONS TO CONSIDER
Question 1 (75%):
Write an essay that reflects pertinent aspects of one of the following essay titles (3000 words)
• Water resources is a political issue and progress will only be made with political support;
• The state of the world water resources is so poor that urgent action should take precedence over the precautionary principle
• Local action in the IWRM is the best option for the poor

- Surface storage reservoirs in IWM: a regional case study
- River basin transfers: a case study
- Conjunctive use of water resources: a case study
- Participation and consultation with community based stakeholders within IWRM: a case study.

(Content-30%; Clarity-25%; Structure-10% and presentation 10%)

THE DUBLIN PRINCIPLES

Essay: Discuss which of the Dublin principles is the most important, in no more than 700 words.

Introduction

The guidelines on how to approach water resources management were reached in Dublin in 1992 during the international conference on water and development. Four key principles came up. These were:

1. Fresh water is a finite and vulnerable resource, essential to sustain life, development and the environment;
2. Water development and management should be based on participatory approach, involving users, planners and policy makers at all levels;
3. Women play a central role in the provision, management and safeguarding of water
4. Water has an economic value in all its competing uses and should be recognised as an economic good.

Relative significance of the Dublin principles

Of the principles, none can be completely disregarded, as all are basic to water resources management and development WEDC, 1994; Jones, 1997). However, central to them all seems to be principle number 2, which states that water development and management should be based on participatory approach, involving users, planners and policy makers at all levels. It seems to form the basis of all the other four. It observes the principle of subsidiarity. This means the people at that lowest decision making level must be fully informed, have full knowledge about the situation, and bear the consequences of their decisions. This is therefore the basic of all the principles. It is most likely to cater for women, the disadvantaged etc. WEDC (2004) state that subsidiarity ensures better design specification, instils a sense of ownership, offers a wider range of design information and more in-depth data, and a better understanding of who the customer is (WEDC, 2004). Further, it guarantees communication and sharing. On the other hand, principle 1 simply restates a reality that water resources are not necessarily permanently renewable, and is the source of life to all its stakeholders. Once it is degraded, it takes time to rejuvenate, and only if it is given time to recover. Connecting this to principle 2, the same stakeholders get involved with participatory approach (Ince and Howard, 1999). Because they use water, stakeholders obviously plan for its use and management. Only that the scale of their involvement has been more of reactive in practice in most places, as opposed to proactive participatory methods proposed by principle 2. Relating principle number 3 to principle 2, women are already covered as stakeholders whose lives are sustained by water, and as front-line participants in development and management of the same. This is critical because women are the majority in most farming and domestic enterprises which demand water, besides being in charge of sanitation, hygiene, ensuring water availability in the household etc. Yet the crucial role played by other stakeholders cannot be underestimated. If there is too much focus on women, dissenting views about them being overworked and overwhelmed with volunteer activities are likely to arise.

Gender equity and gender mainstreaming
Secondly, planning with women at the exclusion of other stakeholders (men) is bound to fail since men are inevitably influential. A win-win option must be applied, and this is exactly what principle 2 does. Thirdly, there are complementary water-related gender roles in most societies, rendering both genders important (WEDC, 2004). Thus planning for successful management and development of sustainable water projects require full gender and stakeholder analysis. The exclusive water development and management efforts are likely to fail from sabotage by neglected quarters. At the moment, almost everything seems to go the women way- and gender issues are increasingly being synonymous with women issues. This is a concept of exclusion, and cannot bring equity to the society. There fore in short, the entire range of stakeholders need to be taken care of in water management and development in a form of partnership, putting special weight on gender issues in water use and development. The principle of comparative advantage can then be applied to assign water development and management tasks to different groups, while being cautious against exclusion of any king- geographical / spatial, social, economic, temporal, cultural, gender etc (WEDC, 2003; WEDC, 2004)). Societies have water-related gender roles.

Water as an economic good: demand driven water management

Lastly, the principle 4 which recognizes water as an economic good cannot be universally applied (Jones, 1997; and WEDC, 2003). This is because any economic good has a price set to control demand and supply, so that equilibrium price prevails. This, however, cannot be the case in a good also with a rights dimension. As an economic good (principle4) water provision will contradict principle 1 which talks about water being a basic necessity of life, and therefore something to which all life is entitled. But if the same basic need is regarded as an economic good, then it creates a vacuum about who to apply which of these two principles to. For one, full application of principle 4 would exclude the poor, while full application of principle 1 would invalidate principle 4. Thus a compromise level must be set fro a balance of both. It means those who can afford can still be given water as a purely economic good, to which they pay for consumption. The economically disadvantaged people can then be given some water simply as a basic need- and they are not to pay- for not only is the water a basic need, but they are also unable to buy it. A tariff schedule which ensures the rich are able to pay indirectly for the water consumed by the poor can help bridge the gap, even though the rich may also view it as infringement of their rights (WEDC, 2003). Thus principle 1 and 4 seem contradictory, while principle 3 is already ably covered by principle 2, as women are as important stakeholder as other stakeholders. This leaves principle 2 as the most important.

Conclusion
Participatory management and development of water resources is therefore the key to sustainable water resources management, and thus the key among the Dublin principles. This approach can guarantee community ownership of any initiative, and therefore sustainability.

CATCHMENT MANAGEMENT PLAN
[2]b: A low-income country aims at starting to prepare catchment management plans. Essay: Using the UK LEAP process as an example, suggest a possible process for preparing the plans, in no more than 1000 words.

Introduction
Integrated Water resources management (IWRM) needs to be done in a coordinated and coherent manner, and in a holistic way catering for the livelihood needs of stakeholders, and using a mix of

---

2

all possible techniques and approaches (Ince and Howard, 1999; and Jones, 1997). In a management, there must be a plan (hereby called management plan) or a strategy. A strategy provides a framework, a direction, and a common point of reference within which more detailed plans and projects can be produced. IWRM plans should be flexible and dynamic to respond to the changing assessments, emerging issues and priorities. This is why a fixed, detailed master plan is not recommended because it does not have the flexibility required of the unpredictable and ever-changing environmental situations. IWRM plans address resource allocation, monitoring, regulation, and enforcement of decisions made. Local Environment agency plan (LEAP) is a case in point (WEDC, 2004).

The LEAP

The LEAP is a typical approach that can be used to manage water resources. This discussion will borrow from the United Kingdom model which is effective, well established and well tested. LEAP is a mechanism that brings the data collection through assessment to plans and actions. This goes through three main steps – including (i) consultation (ii) action plan and (iii) Annual review (WEDC, 2004).

Consultation and public participation (CPP)

The consultation report is an initial working document made as the initial guide, made by a group of facilitators of the LEAP production process. This guides the production of the statement of public consultation (SPC) is a product of wide stakeholder consultation. SPC is a product of exhaustive stakeholder input. A stakeholder analysis needs to be done prior to this process to ascertain inclusiveness. The stakeholders are policy makers (for the enabling environment, setting rules, roles and responsibilities), the Non-governmental organisations (NGOs), the community based organisations (CBOs), Faith based organisations (FBOs), social groups (such as women groups, widow and widower groups, self-help groups, village elder committees, council of elders, clan-based associations, youth groups, merry go round groups, sports and entertainment groups etc), the socially disadvantaged (e.g. orphans, widows, AIDS groups e.g. people living with aids associations, etc), institutions (primary, secondary and nursery schools), research institutions, Government departments, line ministries, International organisations operating in the area, political parties in the area, the local administration (especially the chief, sub chief and village elders), village health workers, farmer associations, cooperative societies, self-help groups, community health workers and community own resource persons, and professional bodies (WEDC, 2003 and WEDC, 2004).

Methods and approaches
The stakeholders of the LEAP may be involved through focus group discussions, written submissions, public gatherings, Key informant interviews, desk review of existing documents, structured and unstructured interviews. The information they need to give include a vision, ambitions, goal, objectives, water allocation, issues affecting each catchment, and proposed solutions, catchment resources and uses, etc (Jones, 1997). This stage needs to be as comprehensive as possible to inform other stages. Participatory methods can be most appropriate here to get as many ideas as possible (WEDC, 2003). Deliberate efforts to include the input of the most disadvantaged are a must if the plan is to be effective. From such a group, salient un-imaginable issues and dimensions are likely to come, which then enrich the whole process and plan.

Production of plan
Step 2 of the LEAP involves the environmental agency (in the Kenyan case, NEMA), drafting a costed program of work to improve and protect the local environment (EMCA, 1999). The NEMA mostly uses its registered lead experts, and this is work the latter can do competently. The drafted program will include different activities, distributed by month and year, and preferably for five years since the Kenyan longer term budget, National and district development plans span 5 years. The program here is fully informed by the consultation results (WEDC, 2004).

Reviews, monitoring and evaluation
The last, but not least, stage involves an annual review. The plan may have been produced, phased and implementation initiated. However, a review may be necessary on regular basis to ascertain that the initial proposals are still valid, given the ever-changing environmental scenarios. Yet, some mechanism must be put in place for annual monitoring and evaluation. Objectively verifiable indicators may have to be identified at the planning stage.

The baseline information collected at stage 1 form the datum against which changes are monitored (Durham and Howard, 2001; and WEDC, 2003). Plan review, monitoring and evaluation identify unforeseen impacts and issues, which then can be addressed before further implementation of potentially impacting activities. A formal project-based environmental impact assessment (EIA) may have to be incorporated as a vital stage 1 of the whole process prior to any proposed project appearing within the wider LEAP (WEDC, 2003). Since the entire product is a plan, its impacts should also be assessed, thereby necessitation a Strategic Environmental Assessment (SEA). Since LEAP is a part of IWRM, and management essentially involves decision making at various levels, decision making is an essential part of the plan. This demands that an analytical strategic environment assessment (ASEA) become a necessity. The stakeholders can be empowered to conduct this by themselves, as a continuing part of consultation and public participation (CPP) (WEDC, 2003; and Durham and Jackson, 2001).

Conclusion
Therefore, a local environment agency plan is an essential aspect of IWRM, and should incorporate the CPP, action planning and annual reviews. The five or ten year review can also be planned for. It needs to be formed at the lowest possible level, observing the principle of subsidiarity, while guaranteeing CPP. This can guarantee sustainable management of resources, water included.

REFERENCES
1. Durham N and Jackson M (2001) People and systems for water, sanitation and health: Monitoring water quality in the developing world. 27th WEDC conference, Lusaka, Zambia
2. EMCA ACT (1999). The Kenya Environment and Coordination agency act. Government of Kenya.
3. Ince M and Howard G (1999) Integrated development for water supply and sanitation: Developing realistic drinking water quality standards. 25th WEDC conference, Addis Ababa, Ethiopia
4. Jones J A (1997). Global Hydrology: Processes, resources and environmental management. Pearson prentice Hall, ISBN 0 582 09861 0. Essex.
5. WEDC (2004) Integrated Water Resources Management (IWRM): A WEDC Postgraduate module. WEDC, Loughborough University, UK.
6. WEDC (2003) Community and management: A WEDC Postgraduate module. WEDC, Loughborough University, UK

Topic task questions and comments on your answers	grade award ed
Question 1	
Write an essay that reflects the pertinent aspects of one of the following essay titles: • Water resources is a political issue and progress will only be made with political support • The state of the world's water resources is so poor that urgent action should take precedence over the precautionary principle • Local action in IWRM is the best option for the poor	C+

- Surface storage reservoirs in integrated water management: a regional case study
- River basin transfers: a case study
- Conjunctive use of water resources: a case study

Participation and consultation with community based stakeholders within integrated water resource management: a case study

Conjunctive use of water resources: a case study of Kasau Village, Lake Victoria Basin, Kenya.

The content of the essay is ok, and it provides an interesting insight into the use of water resources in Kasau village. The presentation could be improved. In particular in this case where you are talking about conjunctive water use you should provide th reader with some idea as to how the various resources are distributed. A map or a sketched diagram would be useful here.

I have difficulty following the structure of the essay. You have a reasonable introduction, that then leads into the description of the various water resources, but it then moves to Lake Victoria, where not only this water resource is introduced, but also modes of transport of water. This could perhaps be better discussed in a separate section. Also water quality considerations are mentioned in relation to Lake Victoria, but these (I would argue) are over-riding arguments that deserve a better place in the hierarchy of the essay.

The table is a reasonable attempt to show the conjunctive use of water in this region.

Explanations are somewhat erratic. For example, when you mention that ponds Okelo, Afulo, and Sitina are seasonal (and by the way indicate that among other source the Okelo is used to supplement these as a drinking water source – confusing…) and that they dry up faster due to the tragedy of the commons syndrome, you really need to explain all of this a bit further. I understand what you're getting at, but it is quite important in the context of current water management and the opportunities offered by conjunctive water use to delve a bit deeper into these issues.

Question 2a	
Discuss which of the Dublin Principles is the most important, in no more than 700 words. The question is tricky – which one is more important when all are so obviously needed to achieve success in IWRM? In essence the four principles read in a logical sequence, firstly recognising the special position fresh water takes in our environment which then leads on to comments on appropriate management and development focusing on the lowest appropriate level of participation and gender awareness and finally leading to a statement on the recognition of water as an economic good. This sequential approach implies perhaps some form of hierarchy with principle 1 arguably the most important. However, without the levels below a hierarchy would not exist and thus interpreting the 'most important' as the one deserving most attention would not lead to optimum management of water resources. Your essay addresses these observations well and concludes with a reasonable choice, given your observations. Well done.	B+
Question 2b	
A low-income country aims to start to prepare catchment management plans. Using the UK LEAP process as an example, suggest a possible process for preparing the plans, in no more than 1,000 words. You've included the process steps of the UK LEAP well enough, but I expected a bit more on how this would work in a low-middle income country. There are some references as to why it is not good to export an established process without due consideration of local conditions. The main different issues raised appear to involve the integration of stakeholders into the	B

process. However, other issues such as the state of the environment, environmental assessment expertise, political climate, priority basis of CMPs, etc. also warrant attention in this context.	

SECTION 6: CASE STUDY

INTRODUCTION:

This stage basically involves identifying and assessing felt needs for improved water and sanitation (WATSAN), as well as health education. It will also involve identification of and topic task of responsibilities, institutional appraisal, demand assessment (DA), demand stimulation (for sanitation and hygiene); stakeholder analysis (SA), accomplishing of technical and social surveys, participatory planning, assessment of existing systems, potential partner identification and partnership formation, preliminary formation of committees, training needs assessment and preliminary training.

GENERAL APPROACH FOR ALL PROGRAM COMPONENTS:

This will take a programmatic, demand responsive approach (DRA) (pp 46 box 3.8, WEDC, 2003), with gender and pro-poor bias. Demand for each component of the program (water, sanitation, hygiene) will be assessed, and where it is low, it will be promoted. The Willingness to pay (WTP) surveys will inform the whole program. For sanitation and hygiene, social marketing (SM) approach will take priority. For both water supply and sanitation, project process guidelines will be adopted (Deverill et al, 2002 pp 38-39). SA will give rise to partnerships formation with all, including financial institutions.

In the whole program, SA, situation analysis, DA, WTP, willingness to charge (WTC), cost-benefit or cost effectiveness analysis (WELL 1998 pp 106-7). DA methods to be adopted include contingent valuation method (CVM) and revealed preference techniques (RPT).Focus group discussions (FGDs) and Key Informant Interviews (KIIs) will support these. RPT will be used to ascertain current expenditure and time spent on WATSAN, while CVM will assess preferences and WTP for new WATSAN options. SM steps will be followed as is given by WELL (1998 pg 210-215).

For the poor-mainstreaming approach to be realised, a Participatory-Ranking-Experience-Perceptions-Partnership (PREPP) approach will be taken using the tools in WEDC (2003 pp 207 box 11.3). Since the program also involves the often low demand sanitation and hygiene, SM to inform and create demand will be necessary, using the PHAST approach (WEDC 2003 pp 208, 219). In forming representative groups, the WEDC (various pp 4) guidelines on social development notes will be followed. Options will be identified from PRA, KII and FGDs. Cost benefit analysis (CBA), Health impact assessment (HIA) and environmental assessment will form part of option selection phase. Operation and maintenance (O & M) will be prioritised.

SPECIFIC OBJECTIVES:
1. Identify a sufficient and wholesome livelihood water supply for the project area;
2. To identify appropriate sanitation arrangements for the inhabitants of the project area;
3. To develop an efficient solid waste management (SWM) system for the project area
4. To develop a strategy for promoting health and hygiene.

SPECIFIC METHODOLOGY:
LIVELIHOOD WATER SUPPLY STRATEGY- PRE-FEASIBILITY STUDY

Task 1: Identify a sufficient and wholesome livelihood water supply for the project area;
This will follow the project preparation approach (WELL 1998 pp 259). The total current and projected water demand will be determined from records and further surveys. Other methods to be followed are as indicated in the general approach above. RPT will be used to ascertain current expenditure and time spent on WATSAN. RPT will be complemented with CVM approach to assess customer preferences WTP for new water options. The procedures for DA will adopt Deverell et al (2002 pp 21) guidelines. During option identification, Deverel et al (2002 pp 18) guidelines will be followed. Emphasis will be on getting the characteristics of each option, followed by developing an effective communication strategy.

Water supply options will follow the criteria given by IRC OP 29E (1995 pp 21 & pp 59). O & M issues be guided by IRC OP 29E (1995 pp 57 and pp 21) where caretaker selection criteria are given. Partnerships will be developed following the guidelines in IRC OP 29E (1995 pp 51-53). For water treatment options, the Social, health, technological, economic, financial, institutional and environmental (SHTEFIE) criteria will be used for option selection (Show 1999 pp 65-67). To provide water for all uses, sectoral water demand assessment will inform design. Baseline surveys will give the number of livestock per village, and FGDs and KIIs will be used to determine variations in livestock numbers and therefore used to project their numbers for planning purposes.

To get the optimal tariff structure, guidelines by IRC OP 29E (1995 pp 71) will be used. Graded rates will be recommended, with options such as minimum tariff (MT), efficiency tariff (ET), environmental efficiency tariff (EET), leakage and environmentally efficient tariff (LEET), and total efficiency tariff (TEF) to be costed and offered to clients in WTP surveys. The institutional options for WATSAN will be borrowed from the six basic management models by WELL (1998 pp 121, pp 124, and box in pp 128). SWOT analysis will be used in institutional appraisal, and will further be augmented with assessment for organizational structures (WELL 1998 pp135), and performance indicators guideline in box 2.6.4 (WELL 1998, pp 136) and IRC OP 29 (1995 pp 75-76).Where possible, Village Level O & M (VLOM) will be emphasized. However, partnership approach will be pursued as a policy to ascertain sustainability.

SANITATION STRATEGY- PRE-FEASIBILITY STUDY
Task 2: To identify the sanitation arrangements for the inhabitants of the project area;
Uptake of sanitation always trails behind water supply because of the former's individual/household nature. This program will adopt the project preparation (WELL 1998 pp 259), as well as project process (Deverell et al, 2002 pp 38-39) approaches. As we strive to generate and inform demand for sanitation and hygiene, extra time, resources and careful planning will be required. Suitability of the treated effluent for use in irrigation will be investigated during effluent treatment options. Thus irrigation water requirements will be determined per crop, with National standards of a hypothetical country to be adopted.

At the stage of option identification, guidelines by Deverel et al (2002 pp 18) will be followed. Emphasis will be on getting the characteristics of each option, then developing an effective communication strategy. For sanitation, which has very strong socio-cultural dimension, a deeper understanding of these two aspects will be necessary during the project cycle. Thus a Participatory Hygiene and sanitation transformation (PHAST) approach will be used to accomplish it. The wastewater disposal options selection will follow the criteria by (WEDC 2003b) pp 1.17. The choice will be based on the following principles: "Best Possible Environmental Option" (BPEO), "Best Available Technology Not Entailing Excessive Costs" (BATNEEC), and will avoid the "something better than Cheapest Available Technology Not Incurring Prosecution" (CATNIP) syndrome (WEDC 2003 b pp 1.17).

SOLID WASTE MANAGEMENT STRATEGY- PRE-FEASIBILITY STUDY
Task 3: To identify and establish an effective system for solid waste management (SWM) strategy.
PRA strategy will be used as per guidelines in WELL (1998 pp 57) to get the baseline. The knowledge, attitude and practice (KAP) gaps will be determined from these preliminary studies so as to inform the program design. Good practice checklists will be designed and used in the analysis. DA using the WTP strategies will be used to gauge costed options. Guidelines by Ali et al(1999 pp 4-5 box 2) will be used, and Ali et al (1999 box 11 pp 17) which gives selection criteria of disposal options.

Each stage of the entire SWM chain will be assessed independently, with source reduction, reuse and recycle (RRR) emphasized at different stages. Since SWM starts from the generation point, which is largely individual or household in nature, demand promotion will be the initial point, even though this will be based on the existing demand level. SWM will be viewed as a special sanitation project, with PHAST procedures followed. These include: SA, PRA, DA, baseline survey, SM and hygiene promotion, adopting the WELL (1998 pp 215) framework on appropriate SWM system.

Proposals for SWM shall follow the rrr policy, with composting, incineration (material recovery) being given priority. This will be incorporated at the option scoring and rating. Vehicles and related equipment selection will be based on; modernity, efficiency and hygienic, durability, efficiency, minimizing handling, labour intensive, existence of local technical capacity, availability of spare parts etc being the key guiding principles. Institutional and revenue generation arrangements will be guided by the polluter pay principle so that SWM charges will be based on toxicity of waste, amount delivered, distance from source, and whether waste already separated or not, using the ranking and scoring method. Tariff guidelines will follow the same guidelines as given under water and sanitation (IRC OP 29E 1995 pp 71)

Task 4: To develop a strategy for promoting health and hygiene.
The SA stage will identify potential partners; SWOT analysis will reveal institutional capacities and limitations (including legislative/ knowledge gaps); while institutional capacity appraisal will reveal area and scope of influence per partner. Based on these, legislation, recruitment and training of staff and an outline of typical information, education and communication campaign strategy will be designed. Guidelines in Deverel et al, (2002 pp 41-42, 44) on preparing a project strategy will be followed. Training will take the SARAR approach (WEDC, 2003 pp 210, 220). The behaviour change on sanitation and hygiene will be accomplished through the PHAST approach, which is effective in helping communities realise behaviour change and to act upon it.

COST RECOVERY STRATEGY IN A WATER PROJECT- PRE-FEASIBILITY STUDY
Supplementary task 1: Suggestions on mechanisms of cost recovery.
Cost recovery strategies will focus on how users can meet the O&M costs, especially after a WTP survey. Mechanisms for payment in kind will be compared. The CAFES principle (Conserving, adequate, fair, enforceable and simple criteria), will be used as a criteria for selection of an appropriate tariff options from among: fixed charge, lifeline block, decreasing block, increasing block, uniform rate, and flat rate (WEDC, 2003 pp 135) and (WELL 1998 pp 114). Tariff formulae options to be used include: minimum tariff, efficiency tariff, environmental efficiency tariff, leakage and environmental efficiency tariff, and total efficiency tariff formulae (IRC OP 29E 1995 pp71-73). On solid wastes, Institutional and revenue generation arrangements will be based on the polluter pay principle so that SWM charges will be based on toxicity of waste, amount delivered, distance from source, and whether waste already separated or not. Payments will be done at delivery based on the above factors. To open up more revenue generation avenues, possibilities of material recovery facilities establishment as part of the program will be assessed.

Expected outcome (PRELIMINARY OPTIONS AND SELECTION CRITERIA)
Objective 1: Identify a sufficient and wholesome livelihood water supply for the project area;
Options to be considered (not exhaustive) are: Pond, river, unprotected springs, streams, lakes, oceans, unprotected wells, boreholes, roof catchment, protected spring, protected

well, protected pond with steps/platforms/ramp/ plinth, Infiltration wells and water tankers options will be used. Choice will be based on reliability, cost, skills and spare parts availability, service level, user preference by gender, and existing and potential water sources (IRC OP 29E 1995, pp 59).

Objective 2: To identify the sanitation arrangements for the inhabitants of the project area;
Options to be considered (not exhaustive) are:
(i) On-site sanitation: Use of bush, simple pit latrine, improved pit latrine, offset pit latrine, ventilated improved pit (VIP) latrine, pour-flush, single pit ventilated, twin pit ventilated, borehole latrine, trench latrine, mobile package latrines, septic tank and aqua privies. Selection criteria will be based on: land availability, method of anal cleansing, upgradability, culture, social aspects, technology, soil permeability, water availability and willingness to pay (Pickford 1991 pp 92-94).
(ii) Off-site sanitation: small bore sewers and conventional sewerage system: Selection criteria will depend on septicity, upgradability, cost, institutional arrangement, and distance from sewer lines.
(iii) Effluent Treatment (ET) options: activated sludge, trickling filter, aerated lagoon, reedbed, biodiscs, waste stabilization pond (WSP) etc. The choice of ET system will depend on cost, land area requirement, effluent volume and quality, system efficiency, desired use of final product, simplicity, sludge treatment and disposal possibilities, aesthetics, power supply, technical and professional support requirements, environmental / cultural/ social considerations, etc (Mara and Pearson, 1998 pp1-4; and WEDC 2003b pp 1.17-8).

Objective 3: To identify and develop an effective solid waste management system for the project area.
The SWM options (not exhaustive) to be considered include: Pit dumping/burning, ground heaping/burning, burying, uncontrolled dumping, incineration, composting, gasification, pyrolysis, separation and reuse, sanitary landfilling. Choice of option will be by criteria in Ali et al, (1999 pp 17, 19-20) and the Integrated solid waste management hierarchy in Tchobanoglous et al (1993 pp 14). More specifically, the choice of option will depend on: pollution (soil, air and water), landfill gas hazard, leachate hazard, resource recovery, handling hazards, upgradability, aesthetics, effects on food chain, odour/health

hazard, volume control, association with accidents and effects on traffic and noise. These will be put in a policy impact matrix and scored or rated accordingly.

Comments on Terms of Reference (ToR)
(i) ToR does not state responsibilities of stakeholders;(ii) There are no timelines.

A brief description of my perception of the study ahead;
Due to the slow pace of sanitation and hygiene uptake compared to water, they will implemented as separate co-ordinated projects, with time and resources allocated to each activity. For water supply, the challenge will be to respond to existing demand. For hygiene, the priority will be identifying and promoting appropriate hygiene behaviour (a KAP study). For sanitation, demand stimulation will be prioritised, as its demand is relatively low before any customer preferred hardware can be laid.

ToR for any further field studies I feel should be undertaken to gather more important information:
1.   Gender analysis (GA) may be useful for effective implementation of the proposed gender mainstreaming (GM).
2.   A 1-5 day GA course may be useful to implementers of the program:
3.   More stakeholder meetings proposed; to take the form of stakeholder steering committee with membership across the board.
4.   This will be informed by a stakeholder analysis (SA) which is also proposed at inception phase to inform degree of influence and importance of each stakeholder. The most influential and important stakeholders will be regularly involved in the project via FGDs or KIIs.This will also form the basis for partnerships.
5.   KIIs and FGDs will continue throughout the feasibility study phase;

Comments on the ToR:
(i) Financial aspects of the expert's work (i.e. payment schedule) are not outlined. (ii) Resources provided by the promoter to the expert are not clear (e.g. office, accommodation, vehicle etc); (iii) Project cost limitations need to be outlined.

Comments on the project information provided (gaps and assumptions)
a) It may be useful to ascertain the radius of influence of borehole 5, instead of relying on estimates (WEDC 2005 pp 13).
b) It may be useful to conduct water analysis of the hand dug wells in the main valley of Nar (WEDC, 2005 pp13).
c) More comprehensive household surveys especially in the underrepresented town of Kadimo may help, as the ones done represented a very small proportion of households. For instance, less than 4% of households in Hariadho were surveyed, while 20% in Ugenya and more than 50% in Acron (WEDC, 2005 pp 32).
CVM useful in capturing the undisclosed information on what the FGD members in Ugenya were WTP (WEDC, 2005 pp 36).

Reporting and Presentation of plans:
The logical framework will be used, and designed using guidelines by WEDC (2003 pp 197) and WELL (2002 pp 304 - 306).

Criteria for success/ Project evaluation
The WHO's minimum evaluation procedure (MEP) (WEDC (2003) pp 14, box 1.7) will be used.

Conclusion:
Indicators of progress in the water and sanitation program will be according to WELL (1998 pp 291) guidelines. To augment the sustainability aspect, the sustainability snapshot as given by WEDC 2003 pp 173 box 9.12 will be used.

REFERENCES:
1. Ali M, Cotton A, and Westlake K (1999) Down to earth Solid waste disposal for low income countries. WEDC, Loughborough, UK.
2. Deverill P, Bibby S, Wedgwood A and Smout I (2002) Designing water supply and sanitation projects to meet demand in rural and peri urban communities. WEDC, Loughborough University, UK.
3. IRC OP 29E (1995) Making your water supply work. IRC occasional paper series 29E. IRC
4. Mara D and Pearson H (1998). Design manual for waste stabilization ponds in mediteranean countries. LTI, Leeds.
5. Pickford J (1991). The worth of water. Intermediate technology publications, London.
6. Tchobanoglous G, Thjeisen H and Vigil S (1993) Integrated solid waste management. Engineering principles and management issues. McGraw Hill Civil engineering series. Singapore.

7. WELL (1998) DFID Guidance manual on water supply and sanitation programs. WELL, Loughborough, UK.
8. WEDC (2005) Hariadho Case Study- project information. WEDC, Loughborough University, UK.
9. WEDC (2003) Community and management. WEDC post-graduate module. WEDC, Loughborough University, UK.
10. WEDC (2003 b). Wastewater treatment. A WEDC post-graduate module. WEDC, Loughborough University, UK.
11. WEDC (various). Integrated water resources management supplementary reading material: Social development notes, note number 39 pp 4: meaningful consultation in environmental assessments (1998).

Comments on the discussion

- Introduction is important but it needs to be very simple. Too many concepts at this early stage may confuse the reader.
- Looking at the background information, are you certain that government officials in the ministry will understand the concepts used, e.g. demand responsive, gender and pro-poor bias, social marketing, etc.?
- Strongly recommend you reduce the number of acronyms used, KII, FGDs, SM, etc.
- What is 'livelihood water supply'?
- Government officials may not be aware of the literature and approaches you are mentioning.
- The topic task shows your understanding of the sector but it may be a useless report for the ministry. Too many acronyms, concepts and literature at this stage devalue your work.

KADIMO WATER SUPPLY AND SANITATION PROJECT PRE-FEASIBILITY STUDY:

Introduction:

This prefeasibility report, written for the Kababa state department of water and sewerage, outlines the preferred solutions for water supply, sanitation, wastewater, solid waste, environmental and hygiene in Kadimo. It is a 20-year WATSAN design for 21,000 people.

IMPORTANT PROBLEMS (OBSTACLES AND CONSTRAINTS) OF THE DESIGN.

Water supply: Assumptions are made on areas where gaps exist e.g. attendance at schools as shown below:

(i) Domestic use: 21,000 people x 100 lcd = 2,100,000 litres = 2100 m^3;

(ii) Public use: city halls, jails, parks, Schools: and dispensaries: 10% of total domestic water demand is used in the design (kapoor, pp 226) = 210 m^3

(iii) Industry: Kadimo also a possible location for meat processing industry (to the East of Nar River) = 30 m^3.

(iv) Total water demand = 2100 + 210 + 30 = 2340 m^3 .

(v) Loss and waste: 15% of total consumption (ie, 15% of i + ii + iii

203

above) = 15% x 2340 = 351 m^3.
(vi) Total design water needs = 2691 m^3.

ASSUMPTIONS:
a. The irrigation effluents do not return to sewers; thus only domestic, industrial, dispensary and school effluents run along sewers.
b. There is an extra daily water demand allowance of (to take care of leakages, loss and waste etc) 15% of total water needs (Kapoor pp 226)
c. Agriculture: livestock and irrigation would need another 100 litres per person per day (assume each person has 3 heads of cattle equivalent)= 21,000 x 3 x 40 lcd =2,520,000 litres/ day = 2,520 m^3/day

AIM OF THE PREFEASIBILITY STUDY:
To select feasible solutions for detailed WATSAN design.

SPECIFIC OBJECTIVES:
1. Identify a sufficient and wholesome livelihood water supply for Kadimo; 2.To identify appropriate sanitation arrangements for the inhabitants of Kadimo; 3.To develop an efficient solid waste management (SWM) system for Kadimo; 4.

OUTLINE OF LIKELY REALISTIC SOLUTIONS BY OBJECTIVE
ITEM 1: WATER SUPPLY:
Objective 1: Identify a sufficient and wholesome livelihood water supply for Kadimo
Hariadho water supply realistic / feasible solutions:
In the longer term, the entire Kadimo will have house tap water supply. The following will apply in the short run for different localities of Kadimo: households with corrugated steel roofs can get drinking water from roof catchment; the houses round a central courtyard can have shared roof water brick tanks for potable water; other water sources like protected spring and wells can be used for other non-drinking water needs; houses separated by low walls can share water sources in yard taps in the short term. Asbestos is eliminated because it is carcinogenic. That residents willingly buy water of unknown quality from vendors at 10 Alod / barrel is a good indication of their willingness to pay. They also pay indirectly for water related diseases' treatment. Vendors will continue being useful in the short term as water suppliers, and in the long run as private water providers and partners. Their water source should, however, be ascertained for users' safety.

The contaminated natural springs and unimproved wells should be protected for short-term use; in longer term, springs can be intake points of potable water. Combined water works intake from both springs and rivers can be the subject of feasibility study in water supply. Shallow wells are eliminated for domestic use in the longer term, because this can lower the water table, are contaminated and can cause subsidence. They can, however, serve livestock. River water is eliminated from among feasible water sources because of dry season contamination. The water supply and design will need to be done in close collaboration with the Kababa-based international NGO with 3000 metre surplus of u-PVC pipes to donate. A donation request should be made to the NGO to help support the abject poor, who can then be supplied with tap water free of charge.

Main and scattered villages: Ugenya, Alego, Sakwa, Gem, and the scattered villages:
The long term domestic use of polluted river Nar is eliminated as it causes water related diseases, but can continue being used for irrigation with care, as well as for domestic use only in very desperate cases, after clarification and disinfection. In the longer term, tap water is the realistic and feasible solution as the villagers are willing to pay. Underground water use can cause salt intrusion and subsidence and is therefore eliminated. The feasibility study will design river and spring intake of water using infiltration gallery to the water works.

The following scenario is proposed in the short run: each school and dispensary to use roof catchment tanks during rains, and treated spring and river water in dry season; and the wells and springs incrementally upgraded and protected for continued use. Proposals will be written to the Danish government funding for rural water supply schemes to dig and develop wells and boreholes in the area, develop the existing unprotected wells and spring, and upgrade the spring as a water works intake in the future. This can take care of part of capital costs, which will then be borne, together with the operation and maintenance costs, by the users. Generally, water users are willing to pay. In the long term, villages should have improved wells or boreholes - which will be designed; in the medium term, they could use yard taps scattered within the school compound as this service is gradually availed. Water from natural unprotected springs and wells is hazardous, and contain pathogens. Disinfection requirements criteria will form part of its design for short term use.

ITEM 2: SANITATION:
Objective 2: To identify the sanitation arrangements for the inhabitants of Kadimo.

(i) On-site sanitation: Use of bush, latrines (simple pit, improved pit, offset pit, ventilated improved (VIP), pour-flush, single pit ventilated, twin pit ventilated, borehole, trench, mobile package), septic tank and aqua privies. Selection criteria was based on: land availability and land requirements, method of anal cleansing, upgradability, culture, social aspects, technology, soil permeability, water availability and willingness to pay (Pickford 1991 pp 92-94).
(ii) Off-site sanitation: small bore sewers and conventional sewerage system were selected based on septicity, upgradability, cost, institutional arrangement, and distance from sewer lines.
(iii) Effluent Treatment (ET) options: potential solutions: activated sludge, trickling filter, aerated lagoon, reed bed, biodiscs, and waste stabilization pond (WSP). The ET criteria were: cost, land area requirement, effluent volume and quality, system efficiency, desired use of final product, simplicity, sludge treatment and disposal possibilities, aesthetics, power supply, technical and professional support requirements, and environmental / cultural/ social considerations.

Selected options/ Realistic solutions:
Some Kadimo inhabitants use simple pit latrines. The unroofed one cannot be used in rain, and is therefore eliminated. The roofed option is a cheap, most tenable short term to long-term option. Open defecation contaminates surface waters, offers no privacy, has smell and fly menace, and insecure at night. Residents want own managed latrines, are ready to pay for improved services, rendering the project demand driven. Dry latrines are recommended in the short run because of lack of water for water-borne systems, which are recommended in long run. Pour flush latrines and septic tanks are water borne, rendering them untenable in Kadimo in the short run.

The latter, however, are most viable options from the tenth year, due to the distance between villages, which makes it expensive to go for sewerage systems, or its closest counterpart, the small bore sewers. In the longer term, septic tanks with soakaways will be good option for the Kadimo town, and various levels of pit latrines in the villages because they are the cheapest of the septic tank variants, are water borne, are hygienic, upgradable and there is still enough land for soakaway. Other variants of the septic tanks are expensive.

Roofed, child secure, fly and smell-free, improved dry pit latrines with small drop holes and footrests, and made of local materials are non-water borne, safe, secure, cheap, and comfortable, and thus will serve whole families well in the short run. Upgradable non-roofed option will remain for poor families. A target of one latrine per homestead can be set, and the second latrine remains optional where strong cultural beliefs and capacity exist. The project is therefore demand responsive as there is an indication of coping strategies, as well as willingness to pay. The simple improved pit latrine can be good short to long-term design minimum option for all families. For the rich, possibility of moving to medium and long term-upgraded options remain. This can be integrated with the Danish government water supply and sanitation upgrading in the villages outside Kadimo town.

The short-term realistic solutions therefore include improved pit latrine and pour-flush latrine. In the medium term, the simple pit latrines can be upgraded to VIP. In the longer term, water seal pour-flush latrines and septic tanks will be appropriate for the villages, schools and dispensaries. Septic tanks can also be upgraded to small-bore sewers, unlike simple pit latrines whose highest upgradable level is VIP. Eventually, all effluents must go to a treatment point, the WSPs. In the longer term, upgradable septic tank eg with small bore sewers, as well as the conventional sewered toilets are the realistic solutions as there will be water; they pose no environmental, health, safety and nuisance hazards; and have low running costs.

ITEM 3: EFFLUENT MANAGEMENT OPTIONS / ALTERNATIVES
Objective 3: To identify the domestic effluent arrangements for Kadimo
The potential solutions were: activated sludge, aerated lagoons, trickling filter and WSP system. the WSP is considered a most realistic alternative than the rest because of the following: lower energy requirement for pumping; lower Capital cost; less miscellaneous energy costs; less labour costs; less Cost of imported items/ forex; less operation and maintenance costs; high Pathogen removal; superior Protozoa cyst and helminthes removal; General suitability and appropriateness for the hot area; higher Technical feasibility; low Smell hazard with good design;

In the rating and scoring result, the waste stabilization ponds gets a 57/60 (95%); followed by 48% (19/40) for aerated lagoons; 21/60

(35%) for activated sludge; and 17/40 (43%) for trickling filter. These scores rate WSPs first, followed by aerated lagoons, then trickling filters, and lastly activated sludge. WSPs are cheap to construct and maintain, and have low energy requirements, thus qualifying as the realistic solution. It heavily consumes land, but this is a sparsely populated rural town. 2-3 maturation ponds are recommended for detailed design, with the final effluent being recycled for irrigation, toilet flushing, and fish farming. The quantity of effluent = 75% of water needs = 75% x 2691 = 2018.25 m^3 $\approx$ 2020 m^3 / day.

SOLID WASTE MANAGEMENT (SWM)
Objective 4: To identify an effective solid waste management (SWM) system.
Introduction:
The SWM potential solutions were: Pit dumping/burning, ground heaping/burning, burying, uncontrolled dumping, incineration, composting, gasification, pyrolysis, separation and reuse, sanitary landfilling. The elimination criteria were: hazard potential (pollution, landfill gas, accident, handling, health, traffic, odour, noise, and leachate), resource recovery, upgradability, aesthetics and volume control. These were be put in a policy impact matrix, scored and rated. The options are compared below, with a realistic solution reached. Incineration is not considered because ToR considers it an option of last resort, while gasification and pyrolysis are not technologically feasible.

Storage: Option of communal and primary storage will be analyzed. Primary storage seems to be better because it is in harmony with the proposed separation and composting, is cheaper, and puts households directly responsible. Fixed containers, fixed storage bins, depots and demountables will be considered. Depots are most appropriate for town area- so will be designed for Kadimo town with high population density. In the villages, however, depots will be given priority because they can be made from local materials, and keep wastes dry. Demountables need vehicles to operate, so are eliminated.

Collection methods: Hand powered options e.g. use of handcarts and animal powered vehicles (carts) in town will be designed because they help build local capacity, provide employment, and involve the community. Vehicles and related equipment selection will be based on: modernity, efficiency and hygiene, durability, efficiency, minimizing handling, labour intensive, existence of local technical capacity, availability of spare parts etc being the key guiding principles. The possibility of applying for one 10-ton truck for use in Kadimo will be recommended in the design.

Solid Waste Treatment and disposal Options:
i. Heap and burn within own compound (ii) give tender to private firm to collect and take away
iii. Construct own water-proof shed for storing, sorting, selling / reusing/ recycling

SOLID WASTE MANAGEMENT OPTIONS / ALTERNATIVES ANALYSIS

Solid Waste Management Option	Heap + burn Score/5	Give tender to private firm score/5	Transfer Station/landfill Score/5
Handling hazards	Low5	Varied (high) 2	Varied (Low)4
Collection and transport cost	Low 4	Moderate-high 2	High1
Aesthetic hazard	High1	Low-moderate3	Low5
Energy Saving potential	None1	Minimum (some picking possible) 2	High4
Material recovery potential	None 1	Unknown 2	High 5
Leachate pollution	Low-moderate (ashes) 3	Unknown 2	XNone5
Level of supervision required	Low-none5	Moderate-none4	High (May also be +ve)3
Capital cost	None5	Minimal-none4	High (one-off) 2
Usability in rainy season	Low 2	N/A	High 4
Running cost	Low, but pay casual regular staff4	High Monthly payments) 2	High1
Responsibility by waste generators	Maximum5	Minimum 2	Maximum5
Air pollution hazard	High1	Low-(varied)2	None-low4
Saving of material	None1	Minimum 2	Maximum5
Ground water pollution hazard	High 1	Low3	None5
Final disposal to right site	N/A	medium-Low (May be disposed of elsewhere)2	High5
TOTAL	39/70	32/70	58/75
Score	56	46	77
Rating	2	3	1

The table above shows a comparison among potential SWM solutions, including: (i) no management option- heap and burn; (ii) assign responsibility to private company and (iii) establish a transfer station. According to the rating, the transfer station/landfill option scores 77%. while heap and burn option scores 56%, well below the first option because of water air and aesthetic pollution hazards. The best-rated option is to make a well-ventilated shed with waterproof floor and roofed transfer station with sanitary landfill. Here, all solid wastes are stored, sorted (separated) composted, and sell what is sellable, reuse and recycle some. This gives the waste generator the direct responsibility, saves energy and material, and minimizes pollution-air, water, soils and aesthetic. The transfer station/landfill option will require detailed sampling design, alongside that for material recovery and handling (e.g. composting) facilities as described below.

Designing a transfer station that includes manual handling and recycling facilities;
Transfer station is recommended in the design because it will ensure material recovery. Design will ensure that waste handling, recovery (scavenging), composting, densification (e.g. compacting / bailing etc) and related waste pre-treatment processes are provided for. Its design criteria will be: Capital cost; Current and projected waste generation rate; cost & availability of land (land is large thus will design a split level type); Available technology; Nature of waste; and Running cost.

This will need characterisation of wastes first, and therefore a sampling programme. Design will assess necessity of a convenience transfer station.

Communal composting programme design.
Composting design will form part of the RRR strategy prioritized in the ToR. In composting, residents will be asked to separate wastes at source, support its collection and transport, handling, material recovery and subsequent disposal. The composting design will integrate the following: Availability of market for the compost; Possible uses of compost, including as landfill cover; Availability of composting equipment; technical skills; and quality assurance. Technical assistance and grant support will be sought from the international donors interested in Kadimo.

Solid waste sampling design:
Design will require sampling to facilitate waste characterisation. Sampling design will focus on: Nature of residents; Socio-cultural groups in the area; Climatic conditions; Existing methods of waste management; institutional responsibility for solid waste; local initiatives, including waste diversion processes; Existing policies, laws and regulations relating to solid wastes; Any efforts in solid waste pollution control; Economic characteristics of Kadimo; Hydrological, geological, environmental inventory data; and Population characteristics and dynamics. The design will propose SWM facility upgrading on incremental basis. This is because moving from the current uncontrolled local disposal, and jumping to control, semi-controlled, or even to sanitary landfill system is not sustainable without systems in place.

Designing a landfill site;
A landfill is a fully engineered facility for safe handling and disposing of solid wastes. Design parameters will include: Geology and Hydrology of area; Quantity of waste; Accessibility; Presence and type of cover material; Availability of land; presence of unique endangered flora and fauna; Composition of waste; Distance from service facilities such as airports; Distance from residential area; Reasonable distance from generation point / generators / transfer station; type (e.g. monofill, above ground / land raise etc). Design basis will be incremental upgradability to ascertain sustainability, taking cognizance of the need to gradually develop the project while developing the support capacity in readiness to operate in a

fully engineered disposal facility in the longer run. This would involve increasing rate, level and efficiency of the entire SWM system that eventually feed the proposed landfill.

Task 5: Designing an Abattoir effluent treatment:
At the industrial effluent treatment method selection, guidelines by Deverel et al (2002 pp 18) are followed, as is the case of water. The wastewater disposal options selection have followed the effluent quality, financial, fate of sludge, local nuisance, costs, suitability, land requirement and availability, waste characteristics, power supply, and technological appropriateness.(WEDC 2003b pp 1.17). An anaerobic treatment method is proposed here because: (i) Effluents are too strong- with a BOD of 2100 mg/L; (ii) the effluent is likely to be highly coloured; (iii) Strong smell is likely to come from the effluent (especially oxides of sulphur. The effluent is unlikely to have chemicals toxic to microbes in treatment facilities. In this respect, a twin compartment septic tank with a long retention time will be designed for this strong effluent to improve in quality before being discharged into the public sewer, for subsequent treatment at the waste stabilization ponds. A second alternative for design will be the activated sludge system for industrial wastes.

Task 6: To develop a strategy for promoting health and hygiene.
The PHAST approach, with participatory methods including participatory mapping, interviews, focus group discussions, objective tree and sanitation ladders will inform the design. The sanitation ladder and PHAST seem to have more advantages due to their participatory nature which also observe subsidiarity- a strong aspect of the ToR..

Institutional capacity building, framework and Sustainability:
Partnership programs will be designed, with three-tier maintenance system adopted. The community 3-tier system is selected because it has grassroots support- where at village level, the communities involved in preventive maintenance; there exists a pump minder at ward level, and a skilled district maintenance team. It eliminated the national single tier model and the regional two tier models because they heavily involve the central government and against the principle of subsidiarity.

TARIF SYSTEM

Graded tariff rates (IRC OP 29E (1995 pp 71) are recommended because of the wide household income range, as well as wide sectoral differences. A mix of tariff systems will be used at different levels, depending on the household income and sector. The industrial sector, with high pollution potential, will pay the leakage and environmentally efficient tariff; the scattered village households will pay the minimum tariff; the residents of the four main villages will pay the efficiency tariff, while the poorer within Hariadho will pay the environmental efficiency tariff scale to enable the municipality to collect and treat domestic wastes. A lifeline tariff level will be set for the abject poor whose identity will be received from the village elders. Village Level O & M (VLOM) will be adopted. However, partnership approach will be pursued to ascertain sustainability.

Conclusion:

This report has highlit the entire WATSAN aspects, giving details of options, and conducting the elimination process with a view to getting realistic or feasible solutions. It encourages a conjunctive water use, with the short-term water options being roof catchment, and household treatment of surface, spring and well water. The long-term feasible water supply option is the roof and treated tap water from the river and springs. The realistic solid waste management strategy includes a longer-term sanitary landfill with convenience and transfer stations; with the shorter term solution being streamlining of the system so that there is improved collection, storage and disposal. This should gradually develop towards a sanitary landfill.

The feasible sanitation systems are the simple upgradable latrines in the short run, and water seal / VIP latrines in the longer run. The feasible short term effluent treatment option the upgradable septic tank system, while the longer term options are the waste stabilization ponds served with sewerage, and long retention time septic tank with small bore sewers for abattoir effluents. The feasibility study report will focus on sanitation, waste management, effluent management (domestic and industrial), as well as hygiene promotion.

BRIEF PROPOSED OUTLINE OF THE FEASIBILITY STUDY REPORT
Introduction: Gives a brief on sanitation requirements in Kadimo.
A: SANITATION:
(i)

pour flush latrines,
Design of septic tank, design of small-bore sewers
(ii)
(iv) Addressing community sanitation issues and sustainability: PHAST as a sanitation and hygiene management tool:

B: EFFLUECT TREATMENT FACILITY DESIGN

Total daily wastewater flow (Q m3/d); Suitable slopes for the pond embankments, and a suitable freeboard; Suitable pond shapes and volumes, including depths, lengths and retention times; pond numbers and types.

(i) ANAEROBIC POND DESIGN:

Importance; Design aim; Shape, Preferred length, Pond volume, Mean Hydraulic retention time, Pond depth, pond area at Top Water Level.

(ii) FACULTATIVE POND DESIGN:

Pond area, Freeboard, Pond Length: Width (TWL dimensions), Shape, Preferred length (L) : Width (W) ratio, Surface BOD loading, Pond depth, Retention time, and Pond Dimensions.

(iii) MATURATION PONDS DESIGN:

Rationale; Fecal Coliform removal at 1st maturation pond, at 2nd maturation pond, and at 3rd maturation pond; Maturation pond freeboard, Maturation pond depth, width, and length; Volume of maturation ponds, mid depth area, Lengths and widths of each maturation pond.

C: SOLID WASTE MANAGEMENT DESIGN:

C1: Designing a transfer station that includes manual handling and recycling facilities;
C2: Communal composting programme design.
C3: Solid waste sampling design:
C4: Designing a landfill site;
C5: designing an incinerator for sanitary wastes.
C6: Design of collection system;
C7: collection of storage system.

D: ABBATOIR EFFLUENT TREATMENT DESIGN:

Key design parameters: Number and arrangement of compartments; flow rate, location, dimensions of each compartment (length, width, depth); number of compartments; retention time per compartment, effluent entry and exit depths, top effluent level per compartment, position and dimensions of small bore (diameter, slope and length).

E: DESIGN PROGRAM FOR HYGIENE PROMOTION: Design of hygiene promotion program.

REFERENCES:
1. Ali M, Cotton A, and Westlake K (1999) Down to earth Solid waste disposal for low-income countries. WEDC, Loughborough, UK.
2. Deverill P, Bibby S, Wedgwood A and Smout I (2002) Designing water supply and sanitation projects to meet demand in rural and peri urban communities. WEDC, Loughborough University, UK.
3. IRC OP 29E (1995) Making your water supply work. IRC occasional paper series 29E. IRC
4. KapoorB (1989) Environmental engineering. Khanna Publishers. Delhi
5. Mara D and Pearson H (1998). Design manual for waste stabilization ponds in Mediterranean countries. LTI, Leeds.
6. Pickford J (1991). The worth of water. Intermediate technology publications, London.
7. Tchobanoglous G, Thjeisen H and Vigil S (1993) Integrated solid waste management. Engineering principles and management issues. McGraw Hill Civil engineering series. Singapore.
8. WELL (1998) DFID Guidance manual on water supply and sanitation programs. WELL, Loughborough, UK.
9. WEDC (2005) Hariadho Case Study- project information. WEDC, Loughborough University, UK.
10. WEDC (2003) Community and management. WEDC post-graduate module. WEDC, Loughborough University, UK.
11. WEDC (2003 b). Wastewater treatment. A WEDC post-graduate module. WEDC, Loughborough University, UK.
12. WEDC (various). Integrated water resources management supplementary reading material: Social development notes, note number 39 pp 4: meaningful consultation in environmental assessments (1998).

LONG TERM WATER SUPPLY-BY SOURCE OPTION:

OPTION	Roof catchment		Groundwater/ Borehole		River water		Treated at water works and supplied to users Permanent Spring	
	Remark	Score	Remark	Score	Remark	Score	Remark	Score
Susceptibility to soil erosion	None	5	Low	4	High	5	Medium	3
Discourage luxurious consumption	High	5	Limited	3	Limited	3	None	3
Effect on community members	None to high (+ve) (Reduce flooding and soil erosion)	5	Low (draw down, subsidence)	3	-ve, High (reduce downstream flow)	1	+ve, High (may be tapped easier by members)	5
Likely to save chemical (s) / additives	Highest	5	High	4	Lowest	1	Low	2
Likely to contain runoff / soil erosion	Highest	5	None	1	None	1	None	1
Likely to relieve	High but	5	High	5	High	5	High	5

214

water demand in city	limited							
Susceptibility to air pollution	High	1	None	5	Low	4	Low	4
Salt intrusion hazards	None	5	V High	1	None	5	None	5
Subsidence hazards	None	5	High	1	None	5	None	5
Monthly bills	None	5	None	5	None	5	High	1
Water treatment costs	None	5	Low	4	High	1	None-minimal	5
Quality assurance	High	5	High	5	High	4	Varied / may be unreliable	2
Reliability of supply	Moderate	3	High	5	Medium	3	Low	1
Capital expense	High	2	Moderate	3	High	1	High	1
O& M costs	Lowest	5	Low	4	Moderate	3	Lowest	5
Total score/ 85		69		56		52		44
Rating		1		2		3		4

WASTE WATER (EFFLUENT) MANAGEMENT OPTIONS / ALTERNATIVES ANALYSIS

Category	WSP SYSTEM	Score/5	Aerated Lagoons	Score /5	Activate sludge	Score /5	Trickling filter	Score /5
Energy requirement for pumping KWh/yr	0	5	8 million	2	10 million	1		-
Capital cost	635 million	5	1200 million	3	1678 million	1	1500 million	2
Other energy costs	0	5			36 million	1		
Labour costs	159 million	5			134 million	2		
Cost of imported items/ forex	95.3 million	5			840 million	1		
O& M costs /yr	25.4 million	4			50 million	1		
Pathogen removal	High 99% +	4	80-90	2	80-90	2	80-90	2
Protozoa cyst removal %	100	5	90-99	3	90-99	3	90-99	3
% Helminthes eggs removal	100	5	90-99	3	90-99	3	90-99	3
General appropriateness for the area	Very	5	Low	2	Low	2	Low	2
Technical feasibility	High	5	Low	2	Low	2	Medium	3
Smell hazard with good design	Low	4	Moderate	2	Moderate	2	Moderate	2
Total Score / 60		57/60		19/40		21/60		17/40
Percentage score		95		48		35		43
Rating		1		2		4		3

SOLID WASTE MANAGEMENT OPTIONS / ALTERNATIVES ANALYSIS

Solid Waste Management Option	Heap and burn within compounds	Score	Give tender to private firm to own collect and take away	Score	Construct own water-proof shed for storing, sorting, selling / reusing/ recycling	Score	Heap in pit and bury	Score
Handling hazards	Low	5	Varied (high)**	2	Varied (Low)	4	Low	2
Aesthetic hazard	High	1	Low-moderate	3	Low	5	Low	4
Saves energy	None	1	Minimum (some picking possible)	2	High	4	None	1
Soil pollution hazard	High (kills microbes; Leachate pollution)	1	Low (Leachate)	3	None	5	High	1
Opportunity for intervention	High	4	Minimum (Varies with MoU/ ToR)	2	High	5	High	4
Level of supervision required	Low-none	5	Moderate-none	4	High (May also be positive)	3	Low	5
Capital cost	None	5	Minimal-none	4	High (one-off)	2	Low	4

Running (O&M) cost	Low, but pay casual regular staff	4	High Monthly payments)	2	High	1	Low	4
Level of responsibility by waste generators	Maximum	5	Minimum (NIMBY Syndrome upholds)	2	Maximum	5	Maximum	5
Air pollution hazard	High	1	Low-(varied)	2	None-low	4	None	5
Saving of material	None	1	Minimum (Some scavenging may be possible)	2	Maximum	5	None	1
Ground water pollution hazard	High e.g. from Leachate of ashes	1	Low	3	None	5	High	1
Assurance of final disposal to right site	N/A	-	Low(May be disposed of elsewhere)	2	High	5	High	4
TOTAL/ 70		34/65		33/70		53/70		41/70
% Score		53		47		76		59
Rating		3		4		1		2

Feedback for Case Study

- A better presentation was expected, for example you could include table of contents, improved sectioning etc.
- A more detailed introduction could include something on the context, structure of the report etc.
- It may be more appropriate to explain all the data clearly, so technical and non-technical staff at the client's office can fully understand the report.
- Technical sections are well written and could be further improved by better structured paragraphing, reasoning, e.g. why river Nar has a risk of water related diseases, and quantification of terms such as long/short term etc.
- Is it not too early to assume that villagers will pay for the tap water supply?
- More details have been provided for solid waste management as compared to water and sanitation. Is there a reason for this?
- Referencing needs further improvement.

CASE STUDY 3
KADIMO SANITATION PROJECT FEASIBILITY STUDY:
Introduction:

This feasibility report, written for the Kababa state department of water and sewerage, outlines the preferred solutions for sanitation, wastewater, solid waste, environment and hygiene in Kadimo. It is a 20-year WATSAN design for 21,000 people, of whom 9598 are in Kadimo town; 3329 in the Main villages; and 7876 in smaller villages. It has the following assumptions: Daily total water demand, effluent and solid waste production are 4362 m3; 3030 m3; and 10 tons respectively. Short-term measures cover up to 9 years, while long term solutions start from the tenth year onwards.

Design population by region (population in the 20th year):
Kadimo Town: 4380x $(1.04)^{20}$ = 4380 x 2.19 = 9597.12 = 9598 people; Ugenya, Alego, Sakwa and Gem: 2240 x (1.02)20 = 2240 x 1.486 = 3328.5 = 3329 people; Scattered extended family villages: 5300 x $(1.02)^{20}$ = 5300 x 1.486 =7875.5 = 7876 people. TOTAL PROJECTED (DESIGN) POPULATION OF KADIMO: 20803 (approximated to 21,000 people)

AIM OF THE FEASIBILITY STUDY: To select feasible WATSAN solutions for detailed design.
SPECIFIC OBJECTIVES:
1. To identify appropriate sanitation arrangements for the inhabitants of Kadimo;
2. To develop an effective effluent management system for Kadimo;
3. To develop an efficient solid waste management (SWM) system for Kadimo;
4. To develop a strategy for promoting health and hygiene for Kadimo.

ITEM 1: SANITATION:
Objective 2: To identify the sanitation arrangements for the inhabitants of Kadimo.
(i)On-site sanitation: Use of bush, latrines (simple pit, improved pit, offset pit, ventilated improved (VIP), pour-flush, single pit ventilated, twin pit ventilated, borehole, trench, mobile package), septic tank and aqua privies. Selection criteria was based on: land availability and land requirements, method of anal cleansing, upgradability, culture, social aspects, technology, soil permeability, water availability and willingness to pay (Pickford 1991 pp 92-94).
(ii)Off-site sanitation: small bore sewers and conventional sewerage system were selected based on septicity, upgradability, cost, institutional arrangement, and distance from sewer lines.
(iii) Effluent Treatment (ET) options: potential solutions were: activated sludge, trickling filter, aerated lagoon, reed bed, biodiscs, and waste stabilization pond (WSP). The ET criteria were: cost, land area requirement, effluent volume and quality, system efficiency, desired use of final product, simplicity, sludge treatment and disposal possibilities, aesthetics, power supply, technical and professional support requirements, and environmental / cultural/ social considerations.

Introduction:

217

Some Kadimo inhabitants use simple pit latrines. The unroofed one cannot be used in rain, and is therefore eliminated. The roofed option is a cheap, most tenable short term to long-term option. Open defecation contaminates surface waters, offers no privacy, has smell and fly menace, and insecure at night. Residents want own managed latrines, are ready to pay for improved services, rendering the project demand driven. Dry latrines are recommended in the short run because of lack of water for water-borne systems, which are recommended in long run.

Pour flush latrines and septic tanks are water borne, rendering them untenable in Kadimo in the short run. The latter, however, are most viable options from the tenth year, due to the distance between villages, which makes it expensive to go for sewerage systems, or its closest counterpart, the small bore sewers. In the longer term, septic tanks with soakaway will be good option for the Kadimo town, and various levels of pit latrines in the villages because they are the cheapest of the septic tank variants, are water borne, are hygienic, upgradable and there is still enough land for soakaway. Other variants of the septic tanks are expensive.

Roofed, child secure, fly and smell-free, improved dry pit latrines with small drop holes and footrests, and made of local materials are non-water borne, safe, secure, cheap, and comfortable, and thus will serve whole families well in the short run. Upgradable non-roofed option will remain for poor families. A target of one latrine per homestead can be set, and the second latrine remains optional where strong cultural beliefs and capacity exist.

The project is therefore demand responsive as there is an indication of coping strategies, as well as willingness to pay. The simple improved pit latrine can be good short to long-term design minimum option for all families. For the rich, possibility of medium and long term-upgrading options remains. This can be integrated with the Danish government water supply and sanitation upgrading in the villages outside Kadimo town. The short-term realistic solutions therefore include improved pit latrine and pour-flush latrine. The approach of costed sanitation options in town and rural areas is taken to provide a foundation for demand responsive approach emphasized in this design. It is proposed that the costed options can be used in contingent valuation, alongside them being used to make sanitation ladder charts.

Design of sanitation system: Short term design:

Some Kadimo inhabitants use simple pit latrines. The unroofed one cannot be used in rain, and is therefore eliminated. The roofed simple pit latrine option is a cheap, most tenable short term. The advantage is that it can be upgraded for medium and long long-term use. Kadimo residents want own managed latrines, are ready to pay for improved services, rendering the project demand driven. The simple pit latrines can be used in both in urban and rural parts in the short run. Advantages of simple pit latrine: (i) cheap (ii) Made of local material (iii) the greatest initial indicator of demand for sanitation services (iv) Can be upgraded incrementally to cater for the following disadvantages: (i) Fly menace (ii) smell (iii) child accident.

Design of simple pit latrine and its variants:

Pit size: Designed per family based on volume of wastes expected over the 20 year design period. With a family size of 6.5 persons, @ producing 60g of dry solids. Day-1 (density of 1.3 g/cm^3); 60 g/ person/ day x 6.5 persons x (20years@x366 days) = 927810 g = 928 kg of dry waste / family / year. With a density of 1300 kg/m^3, this comes to 0.7137 m^3. With 20% moisture content, this comes to a total volume of 2.86 m^3/ year. This can be rounded off to 3 m3/ year. This can be increased by 100% to reach 6 m^3, which then caters for unforeseen circumstances, e.g. family size exceeding 6.5, visitors, cleansing material etc. This compares favourable with an alternative design based on the total waste volume expected per person per year of 0.04 m^3/ year, which comes to 4.8 m^3 for a family of 6.5 for 20 years. This can have dimensions of 1.5m x 1m x 4m (depth) OR 2mx 1mx 3m (depth). With the latter design, 2 holes and doors can be made to cater for possibility of two people using the facility simultaneously. The family can choose, depending on its preferences.

Pit cover: Can be wooden planks or logs filled with mud or cow dung; or alternatively with slab. The latter is more costly, but more durable and more amenable to upgrading. The former can rot and sink after 5 years, and is discouraged.

Drop hole: Can be one or two, depending on the pit dimensions. Size should be small enough, measuring not more than 15 cm x 15 cm. This is to make it safe for use by children (from 2.5 years old). Younger children must be given alternatives.

Superstructure: (i) wall: This is the more significant, as it provides privacy. It can be made from posts and fito, and then filled with mud or grass. The problem with grass is snake hazard in this hot place. The door can be made of papyrus reeds. The roof may or may not be built. If done, it can be a simple thatch. This design is cheap, and is likely to cost less than 40,000 Ad even with a simple roof. It can immediately improve the situation for very poor families, and it gives them a chance of incrementally upgrading the structure as economic status or luck improves.

SEPTIC TANK DESIGN:
This comprises prefabricated tanks that serve as combined settling and skimming tanks, as an unheated anaerobic digester, and as a sludge storage tank. A septic tank followed by a soil absorption system constitutes a conventional on-site wastewater management system. This is an upgradable on-site sanitation comprising the following: (i) one or two connected water-tight compartments for receiving and storing the effluents and (ii) a soakaway OR (ii) drainfield.

Septic tank Design components:
Location: At least 100 m from water point e.g. Pipe, tap, surface water, well, borehole etc.
Septic tank design:
  i.Connection from house to deliver domestic effluent into an outside tank;
  ii.A water-tight tank with one or two interconnected compartments, the forts receiving raw sewage, and the second receiving only septage (operational efficiency is not related to number of compartments in the tank);
  iii.A long pipe perforated towards its end to get septage from second tank.
Tank configuration: Most concrete septic tanks are rectangular with a n interior baffle to divide the tank and access ports to permit inspection and cleaning. The larger chamber formed by the interior baffle contains about 2/3 of the tank volume.
Tank material: concrete or fibre glass; the latter is expensive. First tank retains scum and sludge.
Structural integrity: This determines the long term performance. For the proposed concrete tanks, , this must be ensured by the correct placement of reinforcing steel, and proper composition of concrete mixture. This design is based on maximum tank integrity which requires monolithic pouring of walls and bottom of tank, and the top cast in place with reinforcing steel from the walls extending to the top slab, with water seal placed between the wall and the top.
Testing for water tightness: This is necessary for the protection of the environment and for the operation of subsequent processing and / or disposal facilities. . Each tank should be tested for water tightness and structural integrity by completely filling with water

before and after installation. . Hydrostatic testing is done at the factory by filling the tank with water and letting it stand for 24 hours. It is acceptable if no water is lost during this period.

Sizing: Rectangular shape, with minimum capacity of 750 gallons (2.84m3). Depth of 1-1.5 m preferred, with a slant at the bottom towards the first tank to enable sludge to remain there. Two, three and four bed roomed houses require septic tanks of capacities 3.8, 5.7 and 7.6 m3respectively.

Large septic tanks: These are recommended for clusters of homes, institutions and commercial establishments- e.g. the abattoir, bottling factories, and the homes in clusters. These need to be large, with a minimum 5.7 m^3 (1500 gallons). They are designed as plug flow reactors, with the volumetric capacity being equal to about five times the average flow. , taking into account the accumulation of scum and sludge (190 litres/capita/day), and an appropriate peaking factor (about 1.5) which is a factor of safety. Parallel tanks are recommended for institutions to provide redundancy, and to allow for maintenance.

Emptying of septic tanks: Desludging large septic tanks should be after every 3-6 months. This design has assumed 4 months.

Drainfield and soakaway design:

The soakaway is recommended for towns where only small, but still reasonably sizeable plots exist per family, whereas drainfields are for villages, or for communal septic tanks for which exists a lot of space. This can later be upgraded to small bore sewers, which connect to the sewer system. It is best for town where it is relatively reasonable to provide sewerage system because residences are reasonably close. In rural areas, it is not a tenable alternative because the homes are far apart. Thus in villages, on-site sanitation systems will be recommended. The following scenario exists for both cost-wise. Their problem is a possibility of not functioning during rainy season due to high moisture content of soils.

Septic tank costing for Kadimo:

The septic tank- soakaway combination costs a total of 410,000 + 80,000 = 490,000 Kshs, while the septic tank-drainfield option costs 410,000 + 500,000 Kshs = 910,000 Kshs. For communal septic tank (upgraded) discharging contents to sewer (i.e. small bore sewer system), the total cost per family is 90,000 Kshs. The last is the cheapest option, making it most attractive for town residents. For village residents, the possibility of small bore sewers is unavailable, leaving them only with only the first two options of septic tank-soakaway (costing 490,000 Kshs) and the septic tank-drainfield option (costing 910,000 Kshs.). This renders long term rural life more expensive in terms of sanitation.

Improved simple pit latrine: This will have the same pit size and drop hole. However, it can have slab for pit cover (with a provision for vent pipe), and brick wall, with a thatch on the roof. This is more upgradable, and costs about 81,000 Kshs. Improvised foot rests can improve it even more. This can be incrementally upgraded by adding (i) Drop hole covers to reduce fly menace; (ii) placing footrests to make it more comfortable; (iii) replacing thatch with the more durable corrugated sheet (maximum 2 required); (iv) Placing a light-tight door to reduce flies (v) Blocking the roof-wall space to reduce light entry, and therefore control flies (vi) installing a screened vent pipe to control flies and smell. Upgrading can be gradual, and may cost a total of another 20,000 Kshs.

The advantage of the improved simple pit latrine is that it can closely mimic the ventilated improved pit latrine (VIP) with minimum extra expenses. It is recommended as a short-long term and long term option in the low and middle class in the entire Kadimo town and villages. This is because even if water is to be eventually available, they may not afford water-borne sanitation systems, which are four times more expensive. However, if financial support were to be available, and with correct level of demand created for sanitation, possibility of upgrading from option I above to a VIP would be most appropriate for the poor-middle class residents. For the middle class and rich elites, the next alternative is the septic tank. The upgrading options for this class therefore can be (i) Simple - Improved simple - VIP OR (ii) Improved simple - VIP.

COSTED SANITATION OPTIONS IN RURAL AREAS/ VILLAGES:
Short term simple improved pit to long term VIP:
For those in rural areas, who are in scattered residences, they can consider initial cheap but upgradable non-water borne sanitation system- i.e. simple pit latrine, which they can gradually upgrade to VIP latrines for their individual families. This can cost them an initial 60,000Ad (simple, with thatched roof, and slab), then later another 60,000 for incremental upgrading to VIP. This totals 120,000 Kshs. However, regular emptying will be necessary, at 12,000Kshs every 4 years. This gives another 50,000 Kshs, which comes to a total of 170,000Kshs.

VIP for short and long term purposes:

There may be some middle class and rich villagers who can straight away afford a VIP latrine to serve them both in the short run and in the long run, with a possibility of upgrading it to an offset, pour-flush latrine when water becomes available. This option costs an initial 175,000 Kshs, and another 92,000Kshs at upgrading. This option totals 267,000Kshs. However, it will require frequent emptying, costing 12,000Kshs every 4 years. This costs a total of 327,000Kshs, if 50,000Kshs for emptying is included.

Short term humble (non-roofed) pit to long term septic tank with soakaway:
For those in rural areas, who are in scattered residences, they can consider initial cheap non-water borne sanitation system- i.e. humble (non-roofed) pit latrine for their individual families. This can cost them an initial 43,000Kshs. As water becomes available, they can build septic tanks with soakaway, costing them another 410,000 Kshs + 80,000Kshs = 490,000Kshs. This totals 533,000 Kshs. This can be used by middle class villagers who don't want to share septic tanks. Another 50,000 Kshs may be necessary for emptying within the first 4 years.

Short term roofed simple pit to long term septic tank with soakaway:
For those in rural areas in scattered residences, they can consider initial cheap non-water borne sanitation system- i.e. simple roofed pit latrine for their individual families. This can cost them an initial 81,000Kshs. As water becomes available, they can build septic tanks with soakaway, costing them 410,000 Kshs+ 80,000Kshs = 490,000 Kshs. This totals 571,000 Kshs. This can be used by rich villagers who may not like the option of shared septic tanks. Emptying latrine costs another 50,000Kshs, and septic tanks another 80,000, totaling 700,000Kshs.

Institutional options:
Schools, dispensaries etc should initially have simple pit latrines. In the longer term, septic tanks will be appropriate for schools and dispensaries. Eventually, all effluents must go to a treatment point, the WSPs, through either emptying or via sewers. In urban areas, upgradable septic tank e.g. with small bore sewers, as well as the conventional sewered toilets are the realistic solutions as there will be water; they pose no environmental, health, safety and nuisance hazards; and have low running costs. . However, these institutions are far from town and sewer system is not feasible, rendering septic tank with either a soakaway or drainfield the realistic options.

To reduce the need to empty institutional septic tanks frequently, there may be a need to put up drainfields, which are cheaper in the long run for larger populations in rural areas where land is still readily available. For schools, assuming 10 simple pit latrines are established per school, the non-water borne option costs 810,000. Because they get full quickly due to high pupil population, they may need emptying every year, thus making it another 500,000 Kshs.

Thus the Short term sanitation costs in schools or dispensaries is about 810,000 + 500,000 Kshs = 1, 310,000 Kshs in the short run. From the fifth year, the water borne option should have been in place, with large septic tank-soakaway combination. This costs about 200,000Kshs (approximately equivalent to 5 family septic tanks), these are to be emptied every 4 months (Crites and Tchobanoglous), making it thrice per year@ 20,000 Kshs x 5 = 400,000 x 15 years = 6,000,000Kshs. To this is added another 500,000 for drain fields, totaling to 6,500,000Kshs. Thus the Short term sanitation costs in schools or dispensaries is about 1,310,000 Kshs, averaging about 260,000Kshs per year. In the long run, the large septic tank-drainfield option will cost 6,500,000 to the end of design period, making it 433,333Kshs per year. It would therefore be tempting to consider adopting the non-waterborne system in institutions because it is cheaper in the long term. However, the water borne systems offer an extra advantage of instilling a hygienic habit of hand washing among school-going children. This is later discussed under hygiene.

COSTED SANITATION OPTIONS FOR TOWN RESIDENTS.
Short term pour-flush latrine to long term septic tank:
For those in urban areas, and already using pour-flush latrines, they can maintain this but incrementally upgrade to septic tanks with soakaway for their individual families when water becomes available. This can cost them an initial 92,000 Kshs, then 410,000 for septic tank, and another 80,000 for soakaway. However, this requires emptying the latrine and septic tank every 4 years, giving a total 0f about 100,000Kshs. This option therefore costs a total of 682,000Kshs. Because they already have the pour-flush latrines, this is not factored in costing and therefore a total of 590,000Kshs should be sought for this investment. The problem of septic tank without sewer connection is that the soakaway quickly gets full, and is bound to cause drainage problems in the town.

Short term pour-flush latrine to communal long term septic tank:

For those in urban areas, and already using pour-flush latrines, they can maintain this but incrementally upgrade to communal septic tanks connected to sewers when water becomes available. This can cost them an initial 92,000 Kshs for pour-flush latrine, then 90,000 for sewered communal septic tank. This option therefore costs a total of 182,000Kshs. Because they already have the pour-flush latrines, this is needs not be factored in costing and therefore a total of 90,000Kshs should be sought for capital investment, and another 50,000 for latrine drainage within the first 5 years, and 150,000 per family for emptying the septic tank over the 15 years. This is therefore one of the cheapest options available at only 382,000Kshs for the 20 years.

Short term pit to long term communal septic tank-sewered systems: For town residents those who live in enclosed areas, e.g. round central courtyard, possibility of sharing a septic tank in the longer run exists. Because of the likelihood that these may get water earlier, it is strongly recommended they consider communal simple pit latrines (one for 3 families) which demand 30,000 Ad / family to cater for them in the first 4 years. This period may not even require emptying. However, an allowance of two emptying sessions can be factored in, giving about 24,000Kshs. Because these can get full fast, they can plan with communal septic tank discharging to a sewer (one for 8 - 9 families), which cost 90,000 per family, made available to them by the fifth year. The total short to long term cost of this option is 30,000 + 24,000 + 90,000 per family. This totals 144,000 Kshs. The cost of emptying the tank is 900,000 Ad shared by 6 families (50 people). Thus each family will pay an operation and maintenance cost of 150,000. The total is 371,000Kshs per family for the 20 years. This can be the cheapest option in town- initial cheap but presentable non-water borne system, immediately followed by a communal septic tank-sewered system.

Ecological sanitation:
This is a relatively new concept in Kadimo, where literally no sanitation system exists. It can therefore be part of the entire sanitation package for marketing. It is recommended because it emphasizes waste reuse, which is a vital component of the ToR. It will work best in non-water borne systems, rendering it usable from year 1, and preferably in the villages and schools where the labour for this kind of work is likely to exist. However, it is likely to cost a little more than the normal simple pit latrine, estimated at 100,000Ad for purposes of this design. The advantage is that it is bound to bring benefits to the society. It is therefore a suitable

substitute for simple pit latrines. This design recommends that a partnership be developed with schools in Kadimo with a view to selling this idea among pupils and teachers. From here, it is likely to spread faster as these same people are also members of the village.

Sources of funds for sanitation: This can pursue the Danish government funding a WATSAN project, write sanitation proposals and submit to this potential donor for upgrading sanitation in the villages.

Conclusion:
The most expensive is a single family septic tank with soakaway, and should be avoided. The cheapest options involve a straight investment in VIP latrine to serve both short and long term purposes, as well as having a temporary roofless structure with slab, later upgraded to VIP, followed closely by use of shared / communal facilities in the short run, culminating in communal septic tank. This is a highly recommended option in the town.

ITEM 2: WASTE WATER MANAGEMENT OPTIONS / ALTERNATIVES
Objective 3: To identify the domestic effluent arrangements for Kadimo
The potential solutions were: activated sludge, aerated lagoons, trickling filter and WSP system. the WSP is considered a most realistic alternative than the rest because of the following: lower energy requirement for pumping; lower Capital cost; less miscellaneous energy costs; less labour costs; less Cost of imported items/ forex; less operation and maintenance costs; high Pathogen removal; superior Protozoa cyst and helminthes removal; General suitability and appropriateness for the hot area; higher Technical feasibility; low Smell hazard with good design; In the rating and scoring result, the waste stabilization ponds gets a 57/60 (95%); followed by 48% (19/40) for aerated lagoons; 21/60 (35%) for activated sludge; and 17/40 (43%) for trickling filter. These scores rate WSPs first, followed by aerated lagoons, and then trickling filters, and lastly activated sludge.

Lagoons are basically just another form of stabilization ponds, so they need power supply. To avoid duplication between them, only the WSPs will be discussed. Since the trickling filter is not anywhere along the ToR, the activated sludge, whose cost estimates are given in the case study, will be compared with the WSPs. WSPs are cheap to construct and maintain, and have low energy requirements, thus qualifying as the realistic solution. It heavily consumes land, but this is a sparsely populated rural town. 2 maturation ponds are recommended for detailed design, with the final effluent being recycled for irrigation, toilet flushing, and fish farming. The design below gives details of WSP and activated sludge. However, the effluent delivery system is first discussed.

The effluent delivery system:
1. The following are the proposed Components of the Sewerage system
   1. Transmission main of diameter 400mm.
2. Transmission main is 5,000 m long.
3. The gravity sewerage system of uPVC pipes with diameters 400mm gravitating to lift stations and pumping stations
4. Four number Lift stations each equipped with Afridev centrifugal sewage pumps, two driven by electric motors and one driven by a Lister diesel engine. The diesel engine is provided as a standby in the event of power failure. Each lift station discharges via a short length of pressure main into an adjacent manhole from which sewage gravitates.
5. Two pumping stations, each similarly equipped with three Afridev Centrifugal sewage pumps, two driven by electric motors and one driven by a Lister diesel engine
6. A pressure main of diameter 400 mm from each pumping station provided with the necessary air valves. One pumping station pumps into the second one from which sewage is pumped to the sewage treatment works.
7. Industrial pumping station is equipped with two Gorman Rupp centrifugal sewage pumps, both driven by electric motors. Sewage from this pump station is pumped into a lift station.

Maintenance requirements:
There should be byelaws to control quality of effluent that is discharged into the sewers.
The sewage network should be upgraded after 20 years.
Desludging of pump stations should be done once a year either immediately after or before the rains.
Some staff should be trained by those installing the sewerage system on maintenance, health and safety issues etc. records of these should be kept for future reference and troubleshooting.
There should be records of Maintenance and quality control after the transfer of the original staff who were trained by the project.
There should be public awareness programme to educate the public on proper usage of sewerage system.

There should be bypass manholes for diversion of sewage at lift pump stations. These should also exist for the pump stations.

The system should be flushed by the water tankers.

Children should not play along the sewer lines, as they are likely to block the rodding eyes with sand and other play-objects.

Kadimo, being a large village / small urban centre, should have its sewage system managed by a state water and sewerage council / company. This should have sufficient capacity to operate, maintain, expand and even upgrade a system; it should have the authority to also do major works such as upgrading to avoid conflict of duties.

Cost components of the effluent collection and delivery system:

A: Effluent collection:

(i) Estimated cost of system for Kadimo = 640,000 Kshs / household (dwelling).

With a population of 21,000 and household size of 6.5, the number of households = 21,000 / 6.5. This gives 3231 households. If each household requires 640,000 Kshs, total = 640,000 x 3231 = 2.068 billion Kshs.

(ii) Operation and maintenance cost of effluent collection system = 4% x capital cost = 4% x 2.068 billion Kshs = 82.72 million Kshs.

(iii) Capital + running costs = 2150.72 million Kshs

(iv) Cost of imported items = 5% of total cost = 5% of 2150.72 = 107.54 million Kshs

(v) Labour costs: 40% of total cost = 40% x 2150.72 = 860.23 million Kshs.

B: Transmission of effluent:

Capital cost (Cc) = 140,000 x $L^{0.9}$ x D .75. Where: L = Length of the main in metres; and D = diameter of pipe in metres (D> 500mm).

Q = Velocity of flow x area of flow; 2.43 x 10^{-2} m^3/sec = 0.6 m/s x area; area = 2.43 x 10^{-2} m^3/sec/ 0.6 m/s = 0.0405 m^2

Area of flow = 0.0405 m^2. But 0.0405 m^2 occupies only 1/3 of cross sectional area of pipe, then total area of pipe = 0.1215 m.

Since pipe is cylindrical, then assuming radius = r metres, then r^2 = area / 3.142 = 0.1215 / 3.143 = 0.0387; r = 0.1966 = 0.2 m.

Thus the diameter of the pipe (D m) = 0.4 m; Length of main = 500,000 cm = 5,000 m; Assume D = 0.4 m = 400mm (Thus D <500mm).

However, since we do not have an alternative formula for calculating Cc when diameter of pipe < 500 mm, we will use the same formula given above. Thus Cc = 140,000 x $5,000^{0.9}$ x 0.4 .75 = 140,000 x 2133.4 x 0.503 = 1.5023 x 108 = 150.23 mill Kshs.

(C) House connections and basic plumbing and fixtures:

(i) Connecting a house to a street sewer = 130,000 Kshs. This times # of HHs = 130,000 x 3231 HH = 4.2 x 108 Kshs = 420 mill Kshs.

(ii) Cost of basic internal plumbing of a house that is to be connected to the sewer varies between 100,000 and 200,000 Kshs per house. Assuming a figure of 150,000 Kshs / house, this comes to 4.85 x 108 Kshs = 485 million Kshs.

TOTAL Sewage collection, transmission, plumbing and fixtures = (485 +420) +150.23 + (2068 +82.72 +215.72+107.54 +86.23) million Kshs = 3615.44 million Kshs = 3.61544 billion Kshs.

Basic design aspects:

(A) Design population by region:

Hariadho Town: 4380x (1.04)20 = 4380 x 2.19 = 9597.12 = 9598 people @ 100 litres per capita per day (100 lcd) x 9598 = 959800 litres (959.8 m3) per day in Hariadho;

Ugenya, Alego, Sakwa and Gem: 2240 x (1.02)20 = 2240 x 1.486 = 3328.5 = 3329 people @ 100 litres per capita per day (100 lcd) x 3329 = 332900 litres (332.9m3) per day in Ugenya, Alego, Sakwa and Gem;

Scattered extended family villages: 5300 x (1.02)20 = 5300 x 1.486 =7875.5 ≈7876 people @ 100 litres per capita per day (100 lcd) x 7876 = 787600 litres (787.6 m3) per day in Scattered extended family villages.

(B) Total water demand = Domestic + public + industry + Loss and waste water demands.

(i) Domestic use: 21,000 people x 100 lcd = 2,100,000 litres = 2100m^3;

Assumption / policy: provide sufficient water for residents' use but control use to disciplined levels. Thus once the target supply of 100 lcd is reached, efforts will be put to maintain this, while discouraging luxurious use beyond that level.

(ii) Public use: city hall, jail, park, School & dispensary: 10% of total domestic water demand (kapoor, pp 226) = 210 m^3

(iii) Industry: Kadimo considered possible location for meat industry (to the East of Nar) and bottling company = 30+60 m^3 =90m^3.

Therefore total water demand less wastes and losses = 2100 + 210 + 90 = 2400 m^3.

(iv) Loss and waste: 15% of total consumption (i.e., 15% of i + ii + iii above) = 15% x 2400 = 360 m^3.

Total daily water demand =2100 +210 +360m^3 =2670 m^3. Q=75% of water needs =75% x 2670 = 2002.5 m^3 ≈ 2100 m^3 / dy.

(C) Total daily wastewater flow (Q m3/d); Total daily water demand = 2100 + 210 + 360m^3 = 2670 m^3. Q= 75% of water needs = 75% x 2670 = 2002.5 m^3 ≈ 2100 m^3 / day.

(D) Land area requirements: The most land demanding effluent treatment facility is the Waste stabilization pond. This is used as the basic requirement here, regardless of the final technology selected, as it would still give extra land for future development. Mara and Pearson state that design land area is 10m2/ person for WSPs. With a total population of 21,000, this gives 210,000 m^2 = 21 hectares. This is readily available in open, unoccupied parts of Kadimo.

(E): Siting of effluent treatment facility:
The criteria here include;
Maximum slope (where possible) from town and villages towards the plant; (this is impossible since Kadimo town and villages are situated along the lowlands- along the river). However since the river flow is southwards, the facility must be southwards to minimize the lifting demand - and therefore need for lift pumps.;
Minimize pumping / lifting if it must be done: See sentence above;
Shortest sewerage piping possible- to minimize pipe and excavation costs; This must be correlated to the need for at least 500 m away from the closest residence.

Sited in a relatively flat area: thus the area around coordinates (4.5,7), 4.5,9), 5.5, 7) and 5.5, 9). It is immediately south of Kadimo town, west of Alego village and north west of Gem village. This is a relatively flat space, with a maximum slope of 5%, in a wide open area of over 2 km2 (about 200 ha), only 10% of which is needed to be covered by the ponds in the next 20 years. The rest of the land can be spared for future expansion.

Waste Stabilization Pond (WSP) DESIGN:
a) The total daily wastewater flow (Q m3/d):
Total water demand = Domestic + public + industry + Loss and waste water demands.
(i) Domestic use: 21,000 people x 100 lcd = 2,100,000 litres = 2100m^3;
Assumption / policy: provide sufficient water for residents' use but control use to disciplined levels. Thus once the target supply of 100 lcd is reached, efforts will be put to maintain this, while discouraging luxurious use beyond that level.
(ii) Public use: city hall, jail, park, school, dispensary: 10% of total domestic water demand in the design (kapoor, pp 226) = 210 m^3
(iii) Industry: Kadimo considered possible location for meat industry (to the East of Nar) and bottling company=30+60 m^3 =90 m^3.
Therefore total water demand less wastes and losses = 2100 + 210 + 90 = 2400 m³ .
(iv) Loss and waste: 15% of total consumption (i.e., 15% of i + ii + iii above) = 15% x 2400 = 360 m³.

Total daily water demand = 2100+210 + 360m^3 =2670 m^3. Q= 75% of water needs=75% x 2670=2002.5 m^3 ≈ 2100 m^3 d^{-1}.

ASSUMPTIONS:
1. The irrigation effluents do not return to sewers; thus only domestic, industrial, dispensary and school effluents run along sewers.
2. Both factories will be established in Kadimo;
3. Give an extra water demand allowance of (to take care of leakages, loss and waste) 15% of total water needs (Kapoor pp 226);
4. Per capita H2O use/day will be stabilized at 100 m3 even though politicians want a higher figure which they will not sustain. This restriction is to save water through disciplined use, which is in line with waste / effluent reduction policy.
5. Quantity of effluent = 75% of water needs.

b)Slopes for the pond embankments, and freeboard.
Embankment slopes are: (i) Internal: 1:3 (ii) Internal: 1:2
These are the most commonly used design slopes (Mara and Pearson, 1988; WEDC, 2003, 2003b). However, to make them stable, external embankments should be protected from storm water erosion by planting grass to increase stability, as well as provide adequate drainage. Internal embankment protected against erosion by wave action by using a stone riprap at top water level.

c)Suitable pond shapes and volumes, including depths, lengths and retention times.
Three types of ponds will be separately designed. These are: Anaerobic, facultative and Maturation ponds.

ANAEROBIC POND DESIGN:
Importance: Is considered necessary because its inclusion reduces land area required by up to 39% (Mara and Pearson, 1988; WEDC, 2003; 2003b). However, it can normally cause smell if design is appropriate and iff [SO_4^{2-}]>is 500 mg/l. This will be contained by correct design only, as [SO42-] < 500 mg/l. This is necessary because the effluent is strong (BOD>250mg/l) (Cairncross & Feachem, 1983)

Design aim: The aim of this Design will be to minimize smell from anaerobic ponds.
Shape: Rectangular recommended because it minimizes hydraulic short-circuiting.

Prefered length (L) : Width (W) ratio: 3:1 or 2:1. (Mara and Pearson, 1988)

Assuming a similar surface area, choice of L:W ratio would be based on the length of the design diagonal, to minimize short-circuiting, the choice of which is discussed under pond dimensions below. Anaerobic ponds are designed on basis of volumetric BOD loading,? V ($g/m^3.d$) (Mara and Pearson, 1988; WEDC, 2003, 2003b). This is used to determine pond volume.

Anaerobic Pond volume (Va, m^3)

And anaerobic pond volume (Va, m^3) is related to ? v as shown: Va = L1Q/ ? v (Mara and Pearson, 1988, equation 6.2) and (WEDC, 2003b)

Where: Va is the anaerobic pond volume (m^3); Q is the effluent volume in m^3/d; v is the volumetric BOD loading (in $g/m^3.d$); and should range between 100 and 400; and L_1 is the BOD_5 of the effluent.

From Pickford, 1991, pp 41; and table 6.1 (Mara and Pearson, 1988) and WEDC, 2003), at T? 20^0C, ? v = 300 g $O_2/m^3.d$ to provide an adequate factor of safety against odour release for normal domestic effluent with SO_4^{2-} level < 500 mg/l.

Since the design temperature is that of the coldest month is 310C, ? v = 300 g BOD/m3.d. Therefore (? v = 300)

The mean effluent BOD_5 (L1) is calculated from the weighted mean of the different effluents; domestic, abattoir and bottling

Industries using mass balance method as follows:

Assumption: BOD_5 for domestic effluents = 400 mg/l.

Composite L1 = 2100 x 30 + 650 x 60 + 400 x 2100 / 2190 = 63000 + 39000 + 840000/2190 = 942000/2190 =430 mg/l.

Wastewaters with BOD of 430 mg/l are not strong effluents (WEDC, 2003 b). Thus Va = 430 X 2100 / 300 = 3010 m^3.

Thus the anaerobic pond volume (up to top water level (TWL)) is 3010 m3. Va = 3010 m3

Mean Hydraulic retention time (?, days)

From Mara and Pearson, (1988), equation 6.3, Mean Hydraulic retention time (?, days) = Va / Q; Where Q is the total daily wastewater flow (in m3/d); and Va is the volume (m3)of the anaerobic pond. Thus calculated ? = 3010 / 2100 = 1.43 days. According to Mara and Pearson, 1988, the minimum recommended ? for anaerobic ponds is 1 day, though it is common to have 3-5 days. Mara and Pearson (1988) say that at 150C, retention time of 1.5 days is sufficient with wastes of BOD 300mg/L. In our case here, temperatures are higher, but effluents are stronger, with a composite BOD loading of 430 mg/L. To be sure of completely eliminating smells, a retention time of 3 days will be assumed. Thus ? = 3 days. The pond volume (Va) remains 3010 m^3.

Anaerobic Pond depth (Da, m):
If we assume a pond depth of 4.5 metres (ranges between 2 and 5 metres (WEDC, 2003, 2003b; and Mara and Pearson, 1988), we can determine the pond area at the TWL. A higher figure close to the upper limit is chosen because it can help reduce the pond mid depth area (as pond area (Aa) is inversely proportional to pond depth (Da), thus saving land. Therefore Da = 4.5 m.

Pond area at TWL (Aa): (According to (Mara and Pearson (1988))
Mid-depth area of anaerobic pond (Aa) = Va / D =   3010 / 4.5 = 669 m$^2 \approx 670$ m^2.
 Facultative pond area (Af) = 670 m^2.

Freeboard (Ff, m)
According to Mara and Pearson (1988) the Freeboard of a facultative pond (Ff, m) is given by the equation: Ff = (log $_{10}$ Af)$^{0.5}$ - 1;
Therefore F$_f$ = (log 10 670)$^{0.5}$ - 1 = (2.826)$^{0.5}$ - 1 = 1.68 - 1 = 0.68 m. However, according to Mara and Pearson (1988), F$_f$ should not be less than 1.5 m. Thus the minimum figure of 1.5 m is adopted. Thus Freeboard (Ff, m) = 1.5 m

Pond Length : Width (TWL dimensions)
To determine the optimal L:W ratio to accommodate this area, a choice has to be made between a L:W ratio of 3:1 and 2:1. This choice is to be made, with the goal of maximizing the diagonal in mind (which minimizes hydraulic short-circuiting). Determination of whether 3:1 or 2:1 is as shown below.
Assuming 2 designs, A and B, with A having a L:W ratio of 3:1 and design B having L:W ratio of 2:1,
Design:
L:W ratio: 3:1   2:1
Length of diagonal: 3.16a   2.74a
NB: *: 3a^2 = 2b^2; thus a^2 = 2/3 b^2 ; b2 = 3/2 a^2

Therefore design A gives a higher diagonal of 3.16a units compared with design B which gives a 2.74 a units. Thus a L: W ratio of 3:1 will be adopted, as it gives a higher pond diagonal, which is necessary to minimize hydraulic short-circuiting. Thus the

TWL dimensions of the pond adopted are 3a metres for length, and metres for width.

Area (A_f, m^2) of the facultative pond at TWL = $3a^2$ = 670; a^2 = 223 ; a =$\sqrt{223}$ = 14.93 m $\approx$ 15 metres. Therefore the width of the pond (a, m) = 15 m; a = 15 m; 3a =45 m. Thus the design length of fac. pond = 45 m; design width is 15 m.Lf = 45 m; Wf = 15 m.

DIAGONAL OF POND =47.4 M (Distance along which the wind blows).

2 anaerobic ponds can be constructed in parallel, each 45 m X 15 m, so that one is used when the other is in repair, or being maintained (e.g. at desludging). Only one operates at any given time.

FACULTATIVE POND (FP) DESIGN:

Assumptions:

(i)Pond contents completely mixed (ii) No liquid losses by evaporation; (iii) Breakdown of organic matter in pond is represented by 1st order kinetics. (iv) Sufficient land available; (v) Soils have a coefficient of permeability less than 10-7 m/s to avoid the need for pond lining;

Surface BOD loading:

Facultative ponds (FPs) are designed on basis of surface BOD loading (λ_s in kg/ha.d)(Mara, 1976, cited in Mara and Pearson, 1988).

Thus λ_s =   10Li Q/Af (Mara and Pearson, 1988). ....(i)

Therefore Af =10LiQ/ λ_s...(ii)

Where:

λ_s is surface BOD loading in kg/hectare. day. (kg/ha.d); L1 is the wastewater BOD_5 in mg/l; Q is the total daily effluent (waste-water) flow in m^3/day (m3/d); As is the surface area of the facultative pond at top water level (in m2);

λ_s = 350 $(1.107-0.002T)^{T-25}$; ....(iii)

λ_s = 350 $(1.107-0.002\times31)^{31-25}$ = 350 $(1.107-0.062)^6$ = 350 $(1.045)^6$ = 350 x 1.3022 = 455.8 kg/ha.d.Therefore λ_s = 456kg/ha.d

Li is the effluent BOD for the facultative pond, which is the effluent BOD at exit of anaerobic pond. Assuming a 60% BOD removal at over 20°C (Mara and Pearson, 1988), 0.4 of the original BOD is left (assured because of the higher side of retention time of 3 days in the anaerobic ponds). Lf = 430 x 40% = 172

Area of facultative pond:

Af = 10Li Q/ λs = 10 x (430 x 0.4) x 2100/ 456 = 1720x 2100 / 456 = 7924.7 m^2 $\approx$ 0.80 ha. Thus design area of pond is 0.80 ha.

Mean Pond depth (Df,m)
According to Mara and Pearson (1988), a depth of 1.5 m is typical for a facultative pond. This will be the design depth.

Retention time for the facultative Pond (tf):
Retention time for the facultative Pond (λf), in days (d) is given by: tf =Af /Qm, where
Af is the area of the facultative pond in m^2 and; Qm is the total daily wastewater flow, in m^3/day (m^3/d)
(NB: Qm = (Qi + Qe /2); But In this case, Qi = Qe since seepage, evaporation and other losses are assumed to be negligible); thus evaporative losses + seepage losses = 0. Therefore tf = 7924,7 m^2 x 1.5 m/^{2}100 m^3/d = 5.66 days = 6 days.
tf = 6 days.

Similarly, according to Pickford (1991 pp 41), tf = 10Lf Df/ λs; where
Df is the depth of the pond at top water level (TWL), in metres (m); tf Is the retention time of the facultative pond, in days; and Lf is the BOD_5 of facultative pond influent in mg/l; Thus tf = 10Lf Df/ λs = 10 x 172 x 1.5/456 = 5.55days = 6 days.
According to Mara and Pearson (1988), a minimum of 4 days hydraulic retention time is required at temperatures exceeding 20oC. A higher figure of 6 days is within the recommended duration; the longer the retention time, the lower the hydraulic short-circuiting. This stops algal wash out.

Freeboard Fb): Fb = (log10Af)$^{1/2}$ -1..(iv)
(Mara and Pearson, 1988)
Where Fb is the freeboard in metres; and Af is the area of the facultative pond (in m^2)
From equation (iv), Fb = $\sqrt{}$(log 7924.7) -1 = $\sqrt{3.899}$ -1 = 1.975-1 = 0.975m. Thus freeboard of facultative pond=0.98 m = 1.0 m).

Facultative Pond shape:
Rectangular shapes are most common (Mara and Pearson, 1988), and will thus be adopted here because it gives maximum diagonal along which effluents move, minimizing short-circuiting.

Facultative Pond Dimensions (length and width):

Facultative Pond length (Lf) and width (Wf) can be calculated from the area, together with the recommended L: W ratio. According to Mara and Pearson, (1988), L: W for primary facultative pond is 3:1, and 8:1 for secondary facultative pond. Since in this case the WW comes into the facultative pond from an anaerobic pond, the facultative pond is of secondary type since they receive settled wastewater. Therefore pond L: W ratio of 8:1 is used in this design. Thus if pond width = Wm, then L = 8W metres. This gives an area of $8W^2$ m^2 = 7924.7 m^2. Thus $8W^2$ = 7924.7, implying W^2 = 7924.7 / 8 = 990.6. Therefore W = 31.47 $\approx$ 32 m; Length = 8 x 32 = 256 m. Therefore length of the facultative pond is 256 m; the width is 32 m. Lf = 256m; Wf = 32m.

Diagonal of the pond = 258 m. This offers maximum mixing. the minimum distance along which the wind strikes the pond should be 100m, preferably 200m (WEDC (2003)). Thus a figure of 258 metres gives an excellent opportunity for wind mixing the pond contents.

By the time the effluents leave the facultative pond(s), the BOD left is 43 mg/L. Smith (2004) states that a cumulative figure of 90% BOD is removed by the time the effluent leaves the secondary facultative ponds. Thus 10% of 430 mg/L is left = 43 mg/L BOD left. We can, however, deal with a worst case scenario and assume a 85% removal rate, which gives 65 mg/L BOD. If the effluents are to be reused for irrigation, this would still be high. However, even discharge into the nearby river Nar would not be environmentally feasible because according to Mara and Pearson (1988), a BOD level of 50 mg/L is the upper limit for effluent disposal into a surface water body. Since 65 mg/L is high enough, there would be an obvious need for maturation ponds both from the BOD perspective as well as the reduce-reuse-recycle as a ToR and program policy.

However, maturation ponds achieve only a small removal of BOD (25% each), yet their contribution to bacterial and nutrient removal can be significant (Mara and Pearson, 1988). The maturation ponds would reduce BOD as shown below:
Ist maturation pond: 75% of 65 mg/L BOD = 49 mg/L; (Influent 65; effluent 49 mg/L)
2nd maturation pond: 75% of 49 mg/L BOD = 36.6 mg/L BOD; Influent 49; effluent 37 mg/L);
3rd maturation pond: 75% of 36.6 mg/L BOD = 27.4 mg/L BOD. Influent 37 mg/L; effluent 27 mg/L.

Recycling the water obviously requires maturation ponds to reduce the bacterial count to less than 1000FC/100ml, as well as stabilizing the BOD. Thus the decision on whether to have them is YES. But the decision of how many will be determined from bacterial count. This is shown in the maturation pond design section below:

MATURATION PONDS DESIGN:
Maturation ponds are designed on the bacterial removal approach from the effluent; the purpose is to reduce the number of bacteria in the effluent.

WHERE: N_i = # of FC /100ml influent; N_e = # of FC/ 100ml effluent; K_T=1st order rate constant for FC removal, in day-1; ?=Retention time (days)

STAGE 1: FC reduction at the anaerobic pond:
According to Caincross and Feacham, 1993 pp 171, a design value of up to 10^8 is reasonable for N_i. This design will assume a worst case scenario of FC being maximum, i.e., 10^8 / 100ml. Thus if N_i = 1x108 , N_e x $(1+K_T t)$ = 1E8 (where E8 means x 10^8)
ta = 3 days; K_T (at T = 31^0C) = 2.6 X 1.19 $^{T-20}$ = 2.6 X 1.19 ^{11} = 2.6 x 6.78 = 17.62. However, Caincross and Feachem (1993, pp 171) state that anaerobic ponds need a lower rate constant of K/2. Thus Kt = 17.62/2 = 8.81.   Therefore N_e = 1 x 108 / 1 + (8.81 X 3) = 1E8 / 27.43 = 3645643.46. Thus effluent from the anaerobic pond has 3.65 x 106 FC / 100 ml

STAGE 2: FC reduction at the Facultative pond:
Thus N_e x $(1+K_T t)$ = 3.65E6 (where E6 means x 10^6) ; tf = 9 days; N_i = 3.65 x 10^6. K_T (at T = 31^0C) = 2.6 X 1.19 $^{31-20}$ = 2.6 X 1.19 ^{11} = 17.62. Thus N_e = 3.65 x 10^6/ 1 + (17.62 X 6) = 3.65 E6 / 106.72= 34201.7= 34,202 FC / 100 ml.

STAGE 3: FC at maturation ponds:
N_i= 34,202 ; Assume N_e = 1000; (NB: -target minimum fecal coliform count = 1000 / 100 ml, which can be used for unrestricted irrigation (WHO). Let N be the number of maturation ponds (n = 2 or 3); tm is unknown yet; K_T (at T = 31^0C) = 2.6 X 1.19 $^{T-20}$ = 2.6 X 1.19 ^{11} = 17.62. N_e = N_i / (1+ Kt/2 ta) (1+ Kt tf)(1+ Kt tm))n
Assuming tm = 3 (minimum recommended retention time for each maturation pond, regardless of their number)

FC removal at 1st maturation pond;

$Ne = Ni/(1+KT_Tt)$ (If $Ne = ?$; $Ni = 34,202$; $KT = 17.62$; and $t = 3$). $Ne = 34,202/(53.86) = 636$. This is clearly $< 1000 FC/100$ ml. The number of FC after 1st maturation pond = 636 FC/ 100 ml. Since 1000FC/100 ml of effluent is the maximum recommended standard for unrestricted irrigation, 1 maturation pond suffices, which releases effluent with 667 FC/100 ml. However, for purposes of post design period (and in recognition of incremental upgrading of the system), and the need for extra income from the system to cater for the high capital and running costs, a second maturation pond may be planned. It can be used as a fish pond during the design period so that there is extra income from the facility. Later, it can act as an extra treatment pond as the need continues to arise with time.

FC removal at 2nd maturation pond;
$Ne = Ni/(1+K_Tt)$ (If $Ne = ?$; $Ni = 636$; $K_Tt = 52.86$; $K_Tt + 1 = 53.86$). $Ne = 636 / 53.86 = 11.8$ FC/100 ml ≈ 12 FC/100 ml. Since the required FC count standard is already achieved in the 1st maturation pond, this FC count is insignificant. It only indicates that the effluent here is as good as any normal surface water, thus usable for fishing.

FC removal at 3rd maturation pond;
$Ne = Ni/(1+K_Tt)$ (If $Ne = ?$; $Ni = 12$; $K_Tt = 52.86$; $K_Tt + 1 = 53.86$); $Ne = 12 / 53.86 = 0.2$ FC/100 ml $= 1$ FC/100 ml.
Therefore FC count of effluent after passing through 4 ponds (Anaerobic, facultative and 2 maturation in a series), will be 12 / 100 ml. This is excellent, even if the effluent was to be used for drinking, subject to elimination of helminth eggs.

In the short run, having only the first maturation pond is good enough. The effluent quality after maturation pond is good for unrestricted irrigation. After some time, this can be upgraded to 2 maturation ponds, and later to three. This report will assume a design of up to 1 maturation pond up to the 20th year (covering the entire design period). This is reasonable enough because the high temperature makes it possible to achieve reasonable effluent quality with minimum number of ponds. Water from the first maturation pond will be recommended to be diverted for irrigation. The second maturation pond will be designed for establishment as a final resting site for the fish before they are harvested. This can boost income for the water and sewerage company, helping it to recover some of the operation and maintenance costs.

Maturation pond freeboard:

A uniform figure of 1.5 m will be adopted in the design of all maturation ponds. This is a recommendation by Mara and pearson (1988); WEDC (2003 and 2003b). Fb = 1.5 m

Maturation pond depth, width, and length
According to WEDC, 2003b, Surface loading for the maturation ponds (λsm, in kg/ha.d), is given by $\lambda sm = 0.75 \lambda sf$; (where: λsm is the surface loading of the maturation ponds; and λsf is the surface loading of the facultative ponds).
From earlier design, $\lambda sf = 456$ kg/ha.d; Therefore since $\lambda sm = 0.75 \lambda sf$; $\lambda sm = 0.75 \times 456 = 342$ kg/ha.d
According to WEDC, 2003b, hydraulic mean retention time (= 3 days adopted earlier), ; ? = $10 \times Le \times d$ / ?sm
(where: d = depth of pond (in metres); Le = BOD of effluent from facultative pond (mg/l). Therefore, d = t x λsm / 10Le = 3 days x 342 kg/ha.d / 10 x 65 mg/l = 1.5785 m $\approx$ 1.6 metres. Thus pond depth=1.6 m.

Volume of maturation ponds (Vm):
Volume of maturation ponds (Vm) is given by t x Q (where Q is total daily WW flow (M^3/day) and t is the mean hydraulic retention time in days). NB: Total daily water demand = 2100 + 210 + 360m^3 = 2670 m^3. Therefore Quantity of effluent = 75% of water needs = 75% x 2670 = 2002.5 m^3 $\approx$ 2100 m^3 / day. Vm = 3 days x 2100 = 6,300 m^3

Mid depth area of maturation ponds (Am)
According to WEDC (2003b), Am = Q x λm / d = 2100 m^3/day x 3 days / 1.6 metres = 3937.5 m^2. However, according to Mara and Pearson (1988), Am = (2xQx λm) / (2d+(0.001x λm x e)); Where e = evaporative loss. (Assumption: e = 0). Therefore Am = (2x2100x 3) / ((2x1.6) +0)) = 3,937.5 m^2. Therefore Am = 3,937.5 m^2.
Total area of the ponds is given by 3937.5 X 2 = 7875 m^2. Since rectangular shape gives maximum effluent path flow, thus reducing / minimizing hydraulic short-circuiting, it is adopted in the design. The lengths and widths are shown below.

Lengths and widths of each maturation pond:
Length: width = 8:1 (WEDC, 2003b). Therefore 8a2 = 7875; a^2 = 984.38; a = 31.37; m = 32 m. Therefore width of each maturation pond = 256 m. Dimension of each maturation pond = 256 m x 32m. Lm = 256 m; Wm = 32 m

Accommodating and Taking care of treated effluent in wet season:

Since the treated effluent will be used for irrigation and fish farming, complications may arise during rainy season when irrigation is not necessary. Thus there is need for a 5-month capacity pond to store the water during this rainy season. It can guarantee regular irrigation water supply, and improve agricultural productivity of the area, thus helping boost local economy and reduce poverty. This water can be extended to the neighbouring farms if farmers consent. This capacity is of 315,000 m³ of storage per year.

Other aspects of the WSP design:

i.Design with the lowest temperature since the performance of anaerobic ponds increase significantly with temperature. Thus a design temperature has been 31^0C.

ii.Rectangular shape has been adopted in the design of all ponds because it gives maximum path of flow for the effluent, thus minimizing hydraulic short-circuiting,

iii.Vehicular access to all ponds should be possible in all seasons; a distance of 5m is left between any 2 ponds to facilitate this.

iv.If wind direction is seasonally variable, then wind direction in the hot season should be used as this is when thermal stratification is at its greatest. The effluent flow in ponds is in the opposite direction to the wind direction.

v.To facilitate wind-induced mixing in the surface layers, locate the ponds such that the longest dimension (diagonal) lies in the direction of the prevailing wind.

vi.To minimize hydraulic short-circuiting, inlet should be located such that the effluent flows in the pond against the wind direction.

vii. Each pond should have at least 100 (preferably > 200 m) across its diagonal against the wind. Since the maturation ponds have lengths in excess of either of these minima, the angle against the wind is insignificant.

viii.   To cope with lack of storage capacity during wet season when irrigation is unnecessary; it would be useful to establish waste water storage and treatment reservoirs (WSTR) with about five months' storage capacity (i.e. about 315,00 m³.).

At   WSP systems serving more than 10,000 people, it is often sensible to have 2 or more ponds in parallel. The series are equal, and receive the same flow (Mara and Pearson, 1988). This is taken care of in the design. The design allows for operation and maintenance when some gates may have to be closed e.g. during desludging, repair (e.g. leakage sealing) etc. It also caters for earlier commissioning of facultative ponds before the rest to take care of smell. It will have bypass pipe for anaerobic ponds.

Basis of choice of the designs:
1. Simple; (2) Cheap; (3) Facilitates sampling for monitoring; (4) Inlet to anaerobic pond discharge well below the liquid level to minimize short-circuiting; and to reduce scum quantity;

WSP Costing:
WSP Land area requirements: land area requirements for a WSP system is estimated at 10 m^2/ person. Thus with 21,000 people in the entire Kadimo, land area requirement = 21,000 x 10 m^2 = 210,000 m^2 = 21 ha (Mara and Pearson, 1988). This can be got free by the sewerage company as all land belongs to the government, and can be used for such public services.
A: Capital cost (Cc)of WSP facility: Cc = 1.8 x 10^6 x $q^{0.7}$ .Where q = throughput of the plant in m^3/day;
Total daily water demand = 2100 + 210 + 360m^3 = 2670 m^3. However, since 75-80% of water used equals the amount of effluent, the lower figure of 75% is adopted. This design assumes that 70% would be too low.
Thus q= 75% of water needs = 75% x 2670 = 2002.5 m^3 $\approx$ 2100 m^3 / day.
Cc = 1.8 x 106 x $q^{0.7}$ = 1.8 x 10^6 x 2100 $^{0.7}$ = 1.8 x 10^6 x 211.62 = 380.91 x 10^6 = 381 x 10^6 Kshs = 381 million Ad. This includes
(i) 15% component of foreign exchange item =15% x 381 million Kshs = 57.17 million Kshs; and (ii) labour costs amounting to 25%= 25% of 381 million Kshs = 95.25 million Kshs.
B: Annual operating cost (mainly labour cost) = 4% of Cc = 4% of 381 million Kshs = 15.24 million Kshs.
Total cost of WSP (Capital and Operation and maintenance) = 381 million Kshs + 15.24 million Kshs = 396.24 million Kshs.

Activated sludge as Option 2 effluent treatment system:
Costing:
A: Capital cost (Cc)of activated sludge facility: Cc = 13 x 10^6 x $q^{0.58}$ .Where q = throughput of the plant in m^3/day;
Total daily water demand = 2100 + 210 + 360m^3 = 2670 m^3. However, since 75-80% of water used equals the amount of effluent, the lower figure of 75% is adopted. This design assumes that 70% would be too low.
Thus q= 75% of water needs = 75% x 2670 = 2002.5 m^3 $\approx$ 2100 m^3 / day.
Cc = 13 x 10^6 x $q^{0.58}$ = 13 x 106 x 2100 $^{0.58}$ = 13 x 10^6 x 84.506 = 1098.577 x 10^6 = 1.099 x 10^9 Ad $\approx$ 1.1 billion Kshs.
Labour costs = 8% of Cc = 8% of 1.099 billion Kshs = 87.9 million Kshs
Imported items (Foreign exchange component = 50% of Cc = 50% of 1.099 billion Kshs = 549.5 million Kshs.
Energy cost for pumping / lifting effluent (Cp) = 9.6 x 10^6 x Q x H, where Q is the flow through the pump in m^3/s; H is the head across the pump (m). if q =

2100 m³/ day, then Q = 2100 /(24 hrs x 60 mins x 60 sec) =2100 / 86400 m³/s= 0.02431m³/s = 2.431 x 10-2 m³/s

Pumping height/ head (H) =? Assuming the treatment facility is sited in the relatively flat area within coordinates: (4.5,7), 4.5,9), 5.5, 7) and 5.5, 9), then the height difference cannot be more than 5 m. If we assume this maximum, then H = 5.

Therefore Cp) = 9.6 x 10^6 x 10^{-2} m³/s x 5 m Ad =4.8 x 10^5 Kshs = 0.48 million Kshs.

Other energy costs Ce = 12,000 q = 12,000 x 2100m³/day = 25.2 million Kshs.

Other running costs: 3% of capital cost = 3% x Cc = 3% x 1.099 x 10^9 Ad = 32.97 million Kshs

Total cost of Activated sludge = Cc + labour costs + Imported item costs + pumping costs + other energy costs + other running costs = 1099 + 87.9 + 549.5 + 0.48 + 25.2 + 32.97 million Kshs = 1795.05 million Ksahs = 1.8 billion Kshs.

Thus the total cost of WSP and activated sludge are 397 million and 1800 million respectively. Comparing the two, the WSP is recommended as the preferred option. The environmental impacts of the waste stabilization ponds are given in table 3 in the appendix.

Objective 4: To identify an effective solid waste management (SWM) system.

Table 4: Constraints at primary storage stage: (Guidelines adapted from Ali (2003), Ali et al (1999) and Flinton (1976).

CONSTRAINT	How to overcome
Putting wastes in containers	Public education and awareness; Collaborative, participatory methods vital to give credibility to the system
Diversion of use of container (a direct consequence of, and a function of poverty / poverty-related)	Offer public education; Provide waste-related income generating jobs
Poverty / inability to pay for some services e.g. container or door-to door collection service	Design waste-related income generating projects, and have the residents have some control on who works there- with priority given to the most needy
Social & religious constraints	Social- some extra education may be useful to enable such groups bend their rules a bit; Provide services to willing buyers ie some tariff system varied with acceptability of level of service may be necessary;
Waste separation at source	Public education; Pay households for separation service; Provide compost to complying households (those who desire; and an alternative compensation for the rest)
Primary storage in or outside house? (In door to door collection)	Consensus building; need to ask some houses for a compromise especially; Match collection time with after-work hours
Animals tipping wastes in outdoor primary storage	Provide containers with cover; Conduct public education on the need to cover wastes in yards; Propose a loitering domestic animal policy and assign a tax; Public education on domestic animal management and care; Shoot loitering animals on sight, collect and dump
Human scavenging	Provide safer scavenging points e.g. at transfer station; Provide alternative sources of income.
Lack of resources for buying containers	Invoke service charge and provide containers
Theft of containers	Number the containers as per house number; Public education; Form surveillance teams in estates; Privatize service;
Vandalism of facilities	Public education; Form surveillance teams in estates; Privatize services
Decoding appropriate Container size	Conduct a need survey (with diagrams showing design details in the survey); Conduct participatory public workshops.

Primary collection:
This involves transferring the solid wastes from the storage receptacle, door, kerb, etc to the transfer station or disposal point. It therefore involves also emptying the containers. The realistic transport methods are the animal drawn and hand carts. These would transfer the waste to storage receptacle located within a radius of 1 km from where a powered vehicle would be used to transport the waste to the transfer station and subsequently to the disposal point.

Collection method:
Four possible alternatives exist to choose from. These are door to door, block, kerbside and communal. Door to door collection is whereby the crew collects the waste from the house yards into awaiting vehicles outside the yard. It is rare, and is mainly practised in institutions where houses are close to one another. Block collection is whereby the waste is taken from premises by the generators to an awaiting vehicle outside yards. This needs proper timing, as it cannot work if most households have nobody to deliver the wastes. Kerbside collection is whereby wastes are placed in some container or site at the roadside, or at a kerb, awaiting collection. It is common in the roadsides e.g. at bus stops where alighting passengers, pedestrians and travellers waiting for vehicles would deposit their wastes. The most common waste types in kerbside stores are drink containers, food packages and food remains etc. Communal collection method involves the placing wastes at an official communal storage site awaiting collection. It is the most common in residential sites, and a few commercial sites. It always goes along with a communal storage facility.

Of the four collection methods, communal is the cheapest, followed by block, then kerbside, and the most expensive is the house to house. The order also follows an increase in level of service, with door to door being the highest. A scoring and rating comparison among these is given below. A maximum score of 5 is given for best rating, and 1 or 0 for poorest rating. Score of 0 is assigned to door to door collection under social interactions because it is assumed it is a most important aspect in sustaining an integrated solid waste management (ISWMS); it encompasses a software approach which is important in responsibility, management and capacity building at lowest level (subsidiarity). It is assumed in this design that for a long term ISWMS to be established; the community must actively participate in all aspects- socially, technologically, economically etc. Door to door seems to limit things to only economic at the expense of all others. Table below gives rating and scoring table of collection methods.

Table : Scoring and rating among different collection systems

Collection method	Communal score/5	Block score /5	Kerbside score /5	Door to door score /5
Household role in carrying bins	5	4	3	2
Household role in emptying bins	5	-(optional)	1	1
Minimizes waste handling	1	2	3	4
Need for schedule	5	Optional	2	5
Susceptibility to scavenging	1	5	1	4
Crew sizer	2	4	4	1
Creates jobs	2	3	4	5
Complaint on trespassing	5	5	5	1
Level of service	1	2	3	5
Cost of collection	5	4	3	2
Enhances social interractions	4	3	2	0
Amenable to separation at source	1	2	4	5
Total score	37/60	34/50	35/60	35/60
% score	62	68	70	70
Rating	4	3	1	1

Out of the four levels of solid waste collection service, the best rates are the kerbside and door to door collection, scoring 70%, with block collection scoring a close 68%. This implies an insignificant difference among the three highest level of service, and any can be used depending on the convenience. Therefore the discussion has assumed any of the three can be used. Communal collection is cheapest, but seems to score poorly because it does not facilitate separation, proneness to scavenging, aesthetics (sight, smell, flies), and health and safety issues. It would therefore not be appropriate, except with modifications such as: It can be used at level 2 after door to door, kerbside or block collection from willing households has occurred; the collected waste (separated) can be taken by carts to a communal collection point simply as a point of exchanging full and empty colour coded containers. The waiting trailer would have empty containers, which would be taken by the cart crew as they deliver containers full of separated wastes. This is described in more details elsewhere.

Vehicles to use in collection

Primary collection would involve the use of non motorized vehicles because Kadimo is still too small to have wide enough passages for alternative means. However, these may deliver separated wastes to a transfer station/ communal collection point from where motorized vehicles would come for them. The non-motorised means considered here are the hand carts and animal carts.

Animal drawn carts:

These would be useful in rough, sloppy areas. They are also larger than handcarts; they demand no fuel; can go longer distance compared with hand carts (3-4 km); are low cost compared with motor vehicles, produce no noise and the driver can leave the vehicle and assist in loading. However, their use would be less appropriate because (i) slow speed (ii) interfere with traffic (iii) animals have to be taken care of to work, e.g. fed, treated etc. They should take load to a two-level transfer station at which they tip their loads directly into a motorized vehicle at a lower level. This, however, beats the finer and more desirable purpose of a transfer station- picking, sorting etc. Secondly, the handcart would be a new technology in the area; animal training needs skills and it may be difficult in an area where residents are not used to it. The animal power will displace the cheap human labour, thereby contradicting the poverty reduction goal of the project; one animal can easily displace two men from work. Thirdly, an abattoir would create an automatic demand for animal products, thereby rendering animal power less attractive. Thus a hand cart is more recommended in this scenario.

Handcarts:
Condition and standard of use:

The minimum population density for which hand carts could be used is 7200 / km^2, when each km^2 would require 6 collectors and one trailer / transfer point. Surge capacity would not be necessary and thus the ratio of trailers to tractors would be 7:1, 6 trailers at transfer points and one being towed at any given time by one tractor. The use of short range transfer based on handcarts is relevant with: low per capita generation of wastes;

high waste densities; high population density and low wage rates (Flintoff, 1976 pp 67).

Scenario in the entire Kadimo area:
Area = 232 km^2; current population = 12,920; projected population in 20 years = 21,000; current population density = 55.86 (56/ km^2); projected population density =90.5 (91/ km^2). Therefore the use of a handcart may not apply to the entire Kadimo area. It therefore must be analysed by sections e.g. town, major villages and scattered villages to see where handcarts would be relevant and where not.

Table ..: Example of Solid waste management options:

DISPOSAL OPTION	DONE BY	KEY ISSUES / PROBLEMS/ CONCERNS	OPTIONS FOR PROBLEM SOLVING
OPEN AIR BURNING AND / OR PIT BURNING	HOMESTEADS, SCHOOLS, tyre services (Tyres, papers)	Air pollution Drying and defoliation of flora Killing and / or migration of fauna (including endemic species) Extinction of some plant and animal species Fire hazard	Avoid burning
			Take to dumping site
			Increase paper & stationery use efficiency in institutions
			Awareness raising and public education
PIT DUMPING	HOMESTEADS, SCHOOL GARDENS & FARMS	Odor hazards	Compost putrescible waste
		Rodents, flies and vector hazard	Cover waste; education
		Groundwater pollution	Dump poisonous wastes in the designated site
		Surface water pollution during floods	?*3
BURRYING IN SOIL	HOMESTEADS, SOME INSTITUTIONS	Sharps can cause injuries during farming, or may be exposed by burrowing animals e.g. dogs	Dispose of dangerous wastes in designated site
		Soil pollution by non-biodegradable components	Take such wastes to designated places
		Poisoning and killing of vital soil organisms by buried poisons e.g. pesticide residues and containers	Take such wastes to designated places; Hazard awareness training
DISPOSAL SITE DUMPING	INSTITUTIONS, BUSINESS PREMISES,	Noise	Use ear protector
		Dust	Use respirator / full head gear
		Air-borne Litter	Daily picking schedule, install fence
		Road accidents	Introduce some level of control at dumping site
		Air pollution from exhaust	Use unleaded fuel if possible; use transfer stations; ensure payload capacities reached to minimize trips
		Groundwater pollution by Leachate	Incremental upgrading of site

		Health & safety risks to scavengers	Make a transfer station Train and educate the public oh hazards Encourage use of appropriate protective clothing
		Poor aesthetics at site / ugly site	Increase level of Control of waste placement
		Burning of used tyres releases carcinogenic dioxins	Shred, recycle and use for making doormats, mat
	HOSPITALS' INCINERATION RESIDUE	Leachates polluting ground water	Design landfill for special waste (monofill, or general hazardous waste landfill)
		Uncontrolled burning-causes air pollution	Increased institutional control OR privatization
WASTE DISCARDING AT SOURCE AND DUMPING IN UNAUTHORISED SITES	Construction & Demolition	Ugly sites	Use waste as landfill cover material Use to construct roads
	Some Domestic, Saloon and beauty care products,	Health & safety, and fire hazards to neighboring buildings and residents	Dump in special landfill

Option 2: Controlled integrated solid waste management including organized collection and transport, pretreatment / holding station facility and sanitary landfill (ISWMS) . The key issues are discussed below.

Among the realistic SWM options are (i) no management option-heap and burn; (ii) the ISWMS - i.e. an integrated SWM system, with a proper collection, storage systems, a handling facility (a transfer station), and a sanitary landfill. According to the rating, the transfer station/landfill option scores 77%, while no-management (heap and burn/ burry) option scores 56% due to aesthetics and pollution hazards. The best-rated option is to make a well-ventilated shed with waterproof floor and roofed transfer station with sanitary landfill. Here, all solid wastes are stored, sorted (separated) composted, and sell what is sellable, reuse and recycle some. This gives the waste generator the direct responsibility, saves energy and material, and minimizes pollution-air, water, soils and aesthetic. The transfer station/landfill option will require detailed sampling design, alongside that for material recovery and handling (e.g. composting) facilities.

Designing a transfer station that includes manual handling and recycling facilities;

Transfer station is recommended in the ISWMS design because it will ensure material recovery. Design will ensure that waste handling, recovery (scavenging), composting, densification (e.g. compacting / bailing etc) and related waste pre-treatment processes are provided for. All waste from short range / convenience transfer station will be taken to this major facility by tractor-trailer, which will have taken them from communal storage points as brought by carts for subsequent transport to the transfer station To control traffic and to minimize vehicular accidents, no non-motorised vehicle will be allowed in unless with special permission and reason. Most of the waste will already be in the sorted form (from source), and will be recorded at the gate where a weight bridge exists. Here, all wastes will be put together in separate sections depending on the colour code of the container. They will include: Glass section, composting section (for putrescible wastes e.g. food remains); thin plastics (e.g. for wrapping and packaging); thick plastics (e.g. for carrying liquids); paper and cardboard, metals etc. There will also be a section material recovery facility (MRF), and for commingled wastes brought from residences where waste separation is not yet taken up.

At the different sections of the transfer station, separated wastes will be further sorted, packed and sold or taken to a material recovery facility for recycling or other processing e.g. composting. The Department of water and sewerage will manage the facility for one year to see the returns, as well as to organise tender for privatizing it thereafter. They will use the one year to put systems in place in readiness for the private actor.

Type of transfer station:
  Either a level or more complex split level transfer station can be designed.
Option 1: Simple level site station:
Advantages: Is convenient where transfer effected by manually emptying small containers. It is a simple enclosed parking space with a trailer or skip, rendering it cheap. It is also manual, rendering it suitable for Kadimo where labour is cheap. Because there is enough land, it can be with spare space so that (i) the trailer does not have to be reversed into position and (ii) the exchange involves two operations instead of four when the space is limited. It is also upgradable. The disadvantage is that it is more laborious packing wastes into a vehicle- i.e. it purely manual and special and old wastes may pose serious handling hazards. Therefore it is recommended in the shorter run for Kadimo, but can be upgraded to a split level type after 10 years when more waste is being handled, with less density which may need even compressing at such a facility.

Option 2: Split level station:

This is convenient where small vehicles are unloaded directly into large vehicles by gravity. This saves time and reduces contact with waste. Among vehicles which are favoured by this system are <4 m^3 trailer and a tractor; semi trailers of 15 m^3 capacity; rigid open top vehicles with extended sides to provide capacities of > 12 m^3 ; skips of >8 m^3, carried by rigid vehicles of >5-ton capacity; and roll-on containers of >5 m^3, carried by winch-on vehicles. Where waste s are of low density, skips, roll-on containers and semi trailers can be fed by means of a static packer, which compress the wastes inside the body. Due to these many possible uses of the split level station, it seems a better option, and can be used long into the future as the densities of wastes reduce to the level when they need to be compressed. Initially, the facility can be made to be used with tractor-trailer. However, it is expensive. It Initially, it is not advisable, and may be more convenient after about 10 years when larger amounts of waste are handled.

Other structures for the transfer stations:
For populations between 20,000 and 50,000 (which properly covers Kadimo), there may be advantages of building a combined transfer station and a depot. . With the estimate 100 workers employed on refuse collection, street cleansing and ancillary services, the work force is large enough to warrant engaging the services of a site manager. The depot should have the following (i) welfare facilities for workers: lockers, changing rooms, toilets, showers; (ii) small store rooms with brooms, shovels, cleaning materials, lubricants etc; (iii) parking facilities for hand carts for sweepers and refuse collectors; (iv) office and telephone for the manager; (v) a manned entrance with weighing bridge and records book for all people coming in and leaving the site (vi) space for sorting, processing (e.g. composting, packaging, drying), storing, the wastes. There should be fire-fighting gadgets scattered all round the premises (including roof catchment tanks for such emergencies), with more in the waste handling room.

Communal composting programme design.
Composting design will form part of the RRR strategy prioritized in the ToR. In composting, residents will be asked to separate wastes at source, support its collection and transport, handling, material recovery and subsequent disposal. The composting design will integrate the following: assessing availability of market for the compost; possible uses of compost, including as landfill cover; Availability of composting equipment; technical skills; and quality assurance. Technical assistance and grant support will be sought from the international donors interested in Kadimo, especially to

support collection, storage and handling (sorting, separation, composting and recycling technologies). Emphasis will be put on building local capacities (training) on proper handcart and basket design, and public mobilization. A private company will be identified through competitive bidding to manage the sorting, separation, composting and selling of wastes at the transfer station. Windrow composting will be popularized to the residents so that the compost gets market.

The final fate of sorting remnants:
After waste handling at the transfer station, all useful materials have been retrieved for composting, sale, reuse, recycling, etc. The remaining wastes, coupled with products from incinerator, must eventually be deposited somewhere, where they pose minimal, if any, hazards. There are two options to consider here: The engineered and the non-engineered options. The engineered option involves the use of sanitary landfill, whereas the non-engineered options fall much below. They are described below, with a view to shedding light on why the sanitary landfill option is not an immediate priority in Kadimo, and the need for long-term preparation for it by gradually developing a system through the various non-engineered options.

NON- ENGINEERED OPTIONS
This refers to all practices listed above numbered 1-5 in the table below. They are popular because the engineered option requires capital expenditure, a reliable revenue stream and effective primary and secondary collection service. They have high environmental cost, and include fly, mosquito and rodent breeding, water pollution, air pollution from odour and smoke, and degradation of land. This often creates a negative public impression that all land disposal is offensive, leading officials to search for expensive alternatives such as incineration. . With some basic site operations like spreading, compacting and covering, the waste may be contained and some environmental health control may be achieved over burning, fly breeding and waste picking. However, environmental hazards from leachate and gases remain if the site is not fully and properly engineered and managed. This design proposes that the requirements and aspects characterizing options 1 to 5 be seriously considered and incrementally included, as the first steps towards the sanitary landfill, which may come some 20-25 years, i.e., the last phase of the design period. This is because the engineered option is so technically, economically, socially, environmentally, and politically demanding that it requires systematic building of

capacities in these respects before it can be introduced. As at now, Kadimo is still at level 1 and 2 of the options, leaving it many steps behind the sanitary landfill option.

WASTE HIERRARCHY: The following are waste management options available to Kadimo:

Waste management status- options and hierarchy Guidelines

level	Status	Description	Indicators
1	Waste discarded at source	No collection system operates. Waste is deposited by households in streets and open spaces as they generate it	No primary collection; No functional institution responsible for SWM; Scattered waste in streets and open areas; Waste consumption by animals is common; Burning of piles of waste
2	Uncontrolled local disposal	There is primary collection system and waste is taken manually or in carts to a few disposal points. There is no secondary transportation using vehicles. Common in small towns.	There is institutional responsibility for SWM; Waste is from streets to nearby open places; Waste quantities accumulate; Waste picking starts; Waste consumption by animals is common
3	Uncontrolled city / town disposal	Primary and secondary collection is available. Waste is generally removed from the immediate environment and taken in vehicles to undesignated places away from residential areas	There is an institution responsible for SWM; Waste is removed in two stages; Transfer points are provided Often, vehicle drivers decide which disposal point to use; Waste picking continues at all stages.
4	Semi-controlled disposal	Primary and secondary collection is provided. Waste is generally removed from the immediate environment and taken in vehicles to designated places outside the residential areas. There is no management or equipment at the disposal site	Waste disposal options are in planning stage. Vehicle drivers transport the collected Waste to designated sites Waste picking continues at all stages.
5	Controlled disposal	Primary and secondary collection provided Waste is generally taken outside the residential area to designated sites in vehicles. There is some operational control and equipment / plant available at the site, though disposal is not fully engineered.	Engineered disposal options are in the planning stage Vehicle drivers transport the collected waste to designated sites Controls over waste picking at disposal site begins Solid waste authority owns the site Waste picking continues
6	Fully engineered disposal	Waste is disposed of in a fully controlled manner with maximum protection to the environment. This is quite uncommon in low-income countries.	Details of planning and records are available. No waste picking

Source: Ali et al (1999) pp 7

SANITARY LANDFILLING / FULLY ENGINEERED OPTION

Sanitary landfilling is a fully engineered disposal option. It avoids the harmful effects of uncontrolled dumping by spreading, compacting and covering the waste on land that has been carefully engineered before use. Through careful site selection, preparation and management, operations can minimize risks from leachate and gas production both in the present and the future. Site plans and design consider not only waste disposal but aftercare and ultimate land use once the site closes. (Ali et al, 1999). Sanitary landfill is suitable when suitable land is available at an affordable price. This option must be considered after an assurance that pollution could be controlled and human and technical resources are available to operate and manage the site. Ali et al. (1999) state that hazards arising from landfill can vary from one site to another, but depend primarily on a range of factors including waste composition, moisture and climate. As a guideline, sanitary landfilling should meet the minimum requirements as set below:

1   Location:
2   Operation
3   Management and control

Preparation for and Designing of a landfill site;
A sanitary landfill is a fully engineered facility for safe handling and disposing of solid wastes. Design basis will be incremental upgradability of the existing system to ascertain sustainability, taking cognizance of the need to gradually develop the project while developing the support capacity in readiness to operate in a fully engineered disposal facility in the longer run. The intermediate stages would involve establishing and organizing a dumping ground with a transfer station where waste sorting with a view to reduction, reuse and recycling is done. This would run alongside increasing rate, level and efficiency of the entire SWM system that eventually feed the proposed landfill. Eventually, as the wastes become more potent with increasing industrial development, a sanitary landfill will be established, and it will automatically fit into a well running system. When it comes, the following will be the guide on selection of type of landfill.

Selection of landfill type:
There are three main types of landfills according to the state of California (1984) (Tchobanoglous et al 1993). These are: (i) Conventional landfills for commingled MSW (ii) Landfills for milled solid wastes, (iii) Monofill for designated or specialized wastes (iv) Others e.g. (a) maximum gas production system; (b) integrated solid waste treatment units (c) wetland landfills. In places without appropriate cover material, e.g. where all soil is sand, it would require that appropriate material to be imported from elsewhere. To reduce this expense, it would be better to consider seriously the integrated solid waste treatment units. This involves the organic constituents being separated out and placed in a separate landfill where the biodegradation rates would be enhanced by increasing the moisture content of the waste, either by recycling leachates or by seeding with digested

wastewater treatment plant sludge or animal manure. The degraded material would be excavated and used as cover material for new fill areas, and the excavated cell would be filled with the new waste. This can give additional landfill capacity.

These requirements seem too stringent for the humble Kadimo. However, it is the only long term solution, thereby forcing planners to know its requirements and environmental impacts.

Other Salient Features Of A Functional Sanitary Landfill

The development of a workable operating schedule, a filling plan for the placement of solid wastes, landfill operating records and billing information a load inspection plan for hazardous wastes and site safety and security plans are important elements of a landfill operation plan.

Load inspection

The process of unloading the contents of a collection vehicle near the working face or in some designated area, spreading the waste out in a thin layer and visually inspecting the waste to determine whether any hazardous wastes could be present. The presence of a hazardous material can be detected by a hand-held radiation-measuring device or at the weight station. If hazardous wastes are detected, the collecting company is responsible for removing them (polluter pay principle). In some cases, if the company brings such material a second time, heavy fine is levied. If it brings a third time, it is banned from discharging wastes at the landfill. There should therefore be a means of monitoring the quality and quantity of waste at the landfill site.

Landfill closure and post-closure care

These involve what is to happen to a completed landfill in the future. There should be a budgetary provision for maintaining the closed site into perpetuity, mostly 30-50 years into the future. The closure plan must include a design for the landfill cover and the landscaping of the completed site, and long-term plans for controlling runoff, erosion control, gas and Leachate collection and treatment, and environmental monitoring.

Cover and landscape design

These should be a plan to restore the landfill site to its original, if not better state. This is done by beautification procedures (landscaping), which may involve the use of landscape materials such as hard paving (bricks, slabs etc), soft paving (sand, crushed stones etc), waterfront (pond, fountain) or plants (trees, grass, shrubs or ground covers). A combination of these can be used to ensure the site is attractive enough and poses no esthetic pollution. Wherever possible, the scraping and stockpiling of native topsoil for later use as the final cover for the closed landfill is recommended. This is particularly advantageous when the end use is the restoration of the site to its natural condition and native plants are to be used. These need the availability of local soil, which reduces stress factor for plants growing under inherently adverse conditions of a closed landfill site.

Control of landfill gases

Landfill gases must be controlled for as long as they are expected to be generated after the landfill is closed. This may be by use of extraction wells, collector and transmission piping, and gas flaring and/ or combustion facilities. A means of monitoring, collection and management (e.g. by flaring, or re-use) of landfill gas (LFG) should be in the landfill design right at the conception stage. The LFG management should be done throughout the design, construction, use, closure and post-closure stages to avoid fire and air pollution hazards.

Collection and treatment of Leachate

Landfill Leachates are liquids washed from the landfill wastes. They largely comprise organic acids, and have high Biochemical Oxygen demand (BOD) and Chemical oxygen Demand (COD). Leachates therefore have the capacity to contaminate groundwater, but are also able to transport dissolved organic substances that may be released in the unsaturated subsurface environment, by the change in the partial pressure of the constituents in the gas phase (Tchobanoglous et al (1993). There should be a proper Leachate collection and monitoring strategy throughout the entire life of the landfill (i.e. from conception to post-closure). Their characteristics change with age of the landfill, and these changes should be monitored to be sure they are steady. Any major deviation from normal pattern should be studied and monitored more closely, as it may indicate an interaction with other media such as ground water. This is best done by environmental monitoring systems.

Environmental monitoring systems (EMS)

The Environmental monitoring system (EMS) is necessary to ensure that the integrity of the landfill is maintained with respect to the uncontrolled release of any contaminants to the environment. In most cases, the selection of the facilities and procedures to be included in a closure plan depend on the environmental control facilities used during landfill operation before closure. Designers should chose monitoring facilities that can be used to track the movement of any landfill emissions to the water, air and soil environments. This involves regular check on the characteristics of the LFG, leachates and vadose zone. This should cover the monitors in the vadose zone, water wells and well caps, gas probes and survey monuments. Therefore vadose zone monitoring involves both liquids and gases. Liquid monitoring in the vadose zone is done using suction lysimeters.

SLF Post-closure care
The type of care should depend on the use of the site. Closed landfill sites can be used as recreational areas, parks, nature preserves, botanical gardens, crop production, and commercial development. Each use presents its own unique challenge. The facilities at a closed landfill must be maintained over the period of time that the landfill is producing products of decomposition. This ranges mostly from 20-30 years, but 50 is also possible.

Routine inspections: These are done to characterize the condition of the landfill closure facilities.

Infrastructure maintenance: The infrastructure of landfills includes grading and landscape features, drainage control systems, gas management systems, and Leachate control systems (Tchobanoglous et al 1993). This infrastructure must be maintained systematically through a planned schedule of preventive maintenance to protect the integrity of the landfill cover and prevent contamination of air, water and soil environment adjacent to the landfill. Funds and equipment for these activities must be put aside. Special attention should be given to the landfill cap repair, which includes a geomembrane liner. Both runoff and run on surface waters must be controlled. It may be necessary to install and operate storm water pumps after many years of landfill settlement. Maintenance of drainage control systems must be coordinated with maintenance of land surfaces and revegetation of landscape plans

Conclusion.

Going by the diverse nature of wastes generated, no one disposal method can work; instead a mix of incineration for clinical/ infective waste; pit dumping in rural villages and transfer station with ordinary, and later, sanitary landfill would be designed to complement one another. The scattered households in villages would make it uneconomical to collect waste. The approach of first improving the collection system, storage, disposal and other important components of the chain which feed the system, and the gradual incremental upgrading towards a sanitary landfill would be taken.

Concerns, key intervention points, and recommendations:

Due to the extremely demanding requirements of a sanitary landfill, a more workable option should be sought in Kadimo. This can take the form of an ordinary landfill with controls done around it. Scavenging exposes people to serious environmental hazards. Kadimo does not have any material recovery facilities (MRFs); the low material recovery arises from low public awareness of opportunities available in the SWM sector. More players should be involved through public education, vocational training in appropriate technology, separation of wastes at source (especially households) to ease facilitate subsequent processing, so that cheaper products can be available locally. Programmes aiming at changing attitudes are bound to be the most successful, especially if the locals themselves initiate it. Composting can have a very good future if this programme is promoted alongside agriculture because the compost is bound to improve the soil and agricultural income, as well as surplus being available for landfill cover.

Incremental upgrading of the dumping site could reduce this risk of water pollution. Immediate plans to install a sanitary landfill (SLF) may not be sustainable because of technological, technical, institutional, financial, economic, environmental and social deficiencies. Initially, there is need to install a conventional and convenience transfer station. A convenience transfer station is normally built within the landfill facility for safety concerns. It is for unloading of wastes brought to the site by individuals and small quantity haulers. It also serves as a site for recovery of recyclables. Waste is emptied into transfer trailers each of which is hauled to the disposal site, emptied and returned to the transfer station.

Waste picking and scavenging are likely to arise at the dumping ground. There is thus need for an inbuilt Health and safety program and policy to avoid scavenging with bare hands and inappropriate (or no) footwear. Control at the dumpsite will help control traffic and avoid hazards from vehicular accidents, vehicle exhaust fumes, noise, poisoning and infections from the foods, and contagious diseases from contact with infected and toxic wastes. Site control should provide a central site for waste picking. This can be at a safe site within the convenience, and later, conventional transfer station, which can be included in the Sanitary Landfill (SLF) and a legal requirement for all waste handlers, including those visiting any SWM sites, to be in full PPE gear, enforced, followed by credit / financial schemes to acquire this equipment. The waste pickers can be trained, and organized into cooperative societies so that they are more organized, and can operate on a more-business-like, respectful, and formal level. The idea of getting them alternative livelihood may not be practical, as joblessness is rampant in Kadimo. The informal sector (of which they are already a part) can be organized and expanded to cater for more participants in a more organized way. This is the essence of institutional development.

OBJECTIVE 5: DESIGNING AN ABATTOIR EFFLUENT TREATMENT:
At the industrial effluent treatment method selection, guidelines by Deverel et al (2002 pp 18) are followed, as is the case of water. The wastewater disposal options selection have followed the effluent quality, financial, fate of sludge, local nuisance, costs, suitability, land requirement and availability, waste characteristics, power supply, and technological appropriateness (WEDC 2003b pp 1.17). An anaerobic treatment method is proposed here because: (i) Effluents are too strong- with a BOD of 2100 mg/L; (ii) the effluent is likely to be highly coloured; (iii) Strong smell is likely to come from the effluent (especially oxides of sulphur). However, the effluent is unlikely to have chemicals toxic to microbes in treatment facilities. In this respect, a twin compartment septic tank with a long retention time will be designed for this strong effluent to improve in quality before being discharged into the public sewer, for subsequent treatment at the waste stabilization ponds.

E: DESIGN PROGRAM FOR HYGIENE PROMOTION: Design of hygiene promotion program.

The PHAST approach, with participatory methods including participatory mapping, interviews, focus group discussions, and objective tree and sanitation ladders will inform the design. The sanitation ladder and PHAST seem to have more advantages due to their participatory nature which also observe subsidiarity- a strong aspect of the ToR.

Hygiene-sanitation link
Thus the Short-term sanitation costs in schools or dispensaries is about 810,000 + 500,000   Ad = 1, 310,000 Ad in the short run. This costs about 260,000Ad per year. In the long run, the large septic tank-drainfield option will cost 6,500,000 to the end of design period, making it 433,333Ad per year. It would therefore be tempting to consider adopting the non-waterborne system in institutions because it is cheaper in the ling term. However, the water borne systems offer an extra advantage of instilling a habit of hand washing among school-going children. Being the future parents, policy makers and role models in Kadimo, inculcating a culture of hygiene in them while in schools makes it possible to create demand for hygiene much faster, so that between the fifth and tenth year, there will be sufficient demand for, and benefit from hygiene. Children are known to remember messages they get from peers and teachers much longer and always want to implement them personally, as well as having the urge to push the practice in their families. They are therefore likely to be very influential in finally making hygiene and sanitation be entrenched in Kadimo.

FUNDING:
Possible sources of funds:
A: The Kadimo sanitation team will apply to:
(i) The Italian government, which has provided 15 fiat / calabrese rear loading refuse compaction trucks, financed partly by a grant and the rest by a loan, to be considered as a destinations of the one of the other 3 after 12 of these are deployed largely in class 1 cities.
(ii) International development agency of Liechtenstein has arranged funding (70% loan and 30% grant) for 8 hospital waste incinerators with capacities of 200 kg / day to 10 tons per day; locations of these facilities are yet to be determined. Kadimo committee can apply to be considered.
(iii) Seek health and safety training and capacity building support from the Government of Netherlands and the UK, which are jointly funding a study on waste collection and recycling practices in several class 1 conurbations. The comprehensive study will include

considering improving livelihoods of waste-related workers and ensuring better working conditions, especially with regard health and safety issues and child labour.

(iii)   Organise   for   mini-workshopping   from   national representatives who benefit from The Swiss government program involving senior decision makers from national and state governments' visit international waste management exhibition and participated in a study tour to learn about new WW and solid waste management and recycling technologies. This is meant to strengthen their local grip on promotion of waste reduction, reuse and recycling.

OTHER SOURCES OF FUNDS:

(iv)   Apply to the Swiss government program involving senior decision makers from national and state governments' visit international waste management exhibition for technical assistance and grant support on approved projects that develop innovative ways of reducing the amount of solid waste that goes to landfills and that promote composting and recycling.

(v) Collect funds from the users (user pay approach) for sustainability purposes. This means charging the users at various graded tariff levels outlined earlier in this report. They pay for both water and sanitation services, but to be pro-poor, a lifeline tariff would be most useful in guaranteeing nobody is denied a chance for sanitation.

(vi)   Produce material from solid wastes e.g. decomposible organic matter and compose it. Then the product is marketed and sold. This can be a source of employment for the poor and those unable to pay for some of the services, but also a way of encouraging those willing to separate their solid wastes at source as a way of getting income. Even the settled sludge can be used to make marketable products which can be sold as soil conditioners. Instead of the private sector buying or importing cover materials for the landfill, the compost and settled sludge can be sold to them by those producing it. This makes the cost of operation cheaper.

Institutional capacity building, framework and Sustainability:
Partnership programs should be designed, with three-tier maintenance system adopted. The community 3-tier system is selected because it has grassroots support- where at village level, the communities involved in preventive maintenance; there exists a pump minder at ward level, and a skilled district maintenance team.

It eliminated the national single tier model and the regional two tier models because they heavily involve the central government and against the principle of subsidiarity. Graded tariff rates (IRC OP 29E (1995 pp 71) are recommended because of the wide household income range, as well as wide sectoral differences. A mix of tariff systems will be used at different levels, depending on the household income and sector. The industrial sector, with high pollution potential, will pay the leakage and environmentally efficient tariff; the scattered village households will pay the minimum tariff; the residents of the four main villages will pay the efficiency tariff, while the poorer within Hariadho will pay the environmental efficiency tariff scale to enable the municipality to collect and treat domestic wastes. A lifeline tariff level will be set for the abject poor whose identity will be received from the village elders. Village Level O & M (VLOM) will be adopted. However, partnership approach will be pursued to ascertain sustainability. Neighborhood communities can help sort out some difficulties associated with solid waste facilities and infrastructure.

Conclusion:
This report has highlit the entire WATSAN aspects, giving details of options, and conducting the elimination process with a view to getting realistic or feasible solutions. The realistic solid waste management strategy includes a longer-term sanitary landfill with convenience and transfer stations; with the shorter term solution being streamlining of the system so that there is improved collection, storage and disposal. This should gradually develop towards a sanitary landfill. The feasible sanitation systems are the simple upgradable latrines in the short run, and water seal / VIP latrines in the longer run. The feasible short term effluent treatment option the upgradable septic tank system, while the longer term options are the waste stabilization ponds served with sewerage, and long retention time septic tank with small bore sewers for abattoir effluents.

REFERENCES:
1. Ali M, Cotton A, and Westlake K (1999) Down to earth Solid waste disposal for low-income countries. WEDC, Loughborough, UK.
2. Chatterjee A K (1987). Water supply and sanitary engineering. Khanna Publishers, Delhi.
3. Deverill P, Bibby S, Wedgwood A and Smout I (2002) Designing water supply and sanitation projects to meet demand in rural and peri urban communities. WEDC, Loughborough University, UK.
4. IRC OP 29E (1995) Making your water supply work. IRC occasional paper series 29E. IRC
5. Mara D and Pearson H (1998). Design manual for waste stabilization ponds in Mediterranean countries. LTI, Leeds.

6. Pickford J (1991). The worth of water. Intermediate technology publications, London.

7. Tchobanoglous G, Thjeisen H and Vigil S (1993) Integrated solid waste management. Engineering principles and management issues. McGraw Hill Civil engineering series. Singapore.

8. WELL (1998) DFID Guidance manual on water supply and sanitation programs. WELL, Loughborough, UK.

9. WEDC (2005) Hariadho Case Study- project information. WEDC, Loughborough University, UK.

10 WEDC (2003) Community and management. WEDC post-graduate module. WEDC, Loughborough University, UK.

11 WEDC (2003 b). Wastewater treatment. A WEDC post-graduate module. WEDC, Loughborough University, UK.

LOGICAL FRAMEWORK (LOGFRAME) FOR THE SANITATION AND WASTE MANAGEMENT PROJECT IN KADIMO:

Narrative summary	Objectively verifiable indicators (OVI)	Means of verification	Critical Assumptions
SPECIFIC OBJECTIVE: 1. To identify appropriate sanitation arrangements for the inhabitants of Kadimo;			
Narrative summary: access to sanitary facilities.	At least 50% have access to sanitation facility by the 4th year; 75% by 7th year and 100% by 9th year.	surveys, interviews, and observations at household level;	Household economic situation favorable; cooperation among neighbours.
Sanitary facilities owned.	At least 50% own sanitation facility by the 6th year; 75% by 9th year and 100% by 12th year.	surveys, interviews, and observations at household level;	Household economic situation favorable;
Sanitary facilities are used.	: At least 40% use sanitation facilities by 5th year, 60% by 7th year and 90% by 9th year	surveys	Attitudes and practices have changed
sanitary facilities are maintained	At least 80% of existing facilities are in good condition;	spot check during household visits by community health workers; checking for cleanliness of floor;	Residents have maintenance facilities
sanitary facilities are upgraded	30% of households have upgraded sanitary facilities by 10th year, and increases annually by 15% thereafter.	: surveys, interviews, and observations at household level;	Household economic situation favorable;
SPECIFIC OBJECTIVE: 2. To develop an effective effluent management system for Kadimo;			
Narrative summary: access to effluent facilities.	At least 30% have access to effluent facility by the 4th year; 50% by 7th year and 75% by 9th year.	surveys, interviews, and observations at household level;	Household economic situation favorable; cooperation among neighbours.
Effluent facilities owned.	At least 20% own effluent facility by the 6th year; 50% by 9th year and 100% by 12th year.	surveys, interviews, and observations at household level;	Household economic situation favorable;
Effluent facilities are used.	At least 40% use effluent facilities by 5th year, 60% by 7th year and 90% by 9th year	surveys	Attitudes and practices have changed

261

effluent facilities are maintained	At least 80% of existing facilities are in good condition;	spot check during household visits by community health workers; checking for cleanliness of floor;	Residents have maintenance facilities
effluent facilities are upgraded	30% of households have upgraded effluent facilities by 10th year, and increases annually by 20% thereafter.	: surveys, interviews, and observations at household level;	Household economic situation favorable;
SPECIFIC OBJECTIVE 3. To develop an efficient solid waste management (SWM) system for Kadimo;			
Households pay for household collection tins	10% of households pay for bins by 5th year; 30% by 8th year, and 20% increase thereafter.	Household spot check; surveys	Improved household economy
Households own collection bins	20% of households own bins by 5th year, and 30% increase per year thereafter.	Household spot check; surveys	Improved household economy
Households separate wastes	50% of all household with bins separate, and an increase of 40% per year thererafter.	Spot checks	
Waste collection increase	Waste collection increase to 50% by 5th year, to 75 % by 10th year, and to 90% by 15th year	Records check	Improved management
Waste recycling increases	Increased waste recycling reaches 30% by 5th year, and increases by 30% thereafter	records	
4. To develop a strategy for promoting health and hygiene for Kadimo.			
Households own leaky tins and soap	75% of latrines have leaky tins and soap by 9th year	Spot checks during household surveys by community health workers; evidence of use; access to soap and water.	Household economic situation favorable;
Households use leaky tins and soap	80% of all households with leaky tins use them by 10th year, and 95% by 15th year.	Spot checks during household surveys by community health workers; evidence of use; access to soap and water.	Household economic situation favorable;
School pupils influence their family in hygiene awareness	All households with pupils have a high level of hygiene awareness; safe hand washing practice as identified in hygiene strategy, increases by 30% every year among adults and by 40% every year among school going children;	Interviews, spot checks and visits, spot checks during household surveys by community health workers; evidence of use; access to soap and water. observation surveys;	Pupils have a voice in the family

SECTION 7: RESPONSE TO SOME QUESTIONS

Essay: Why solid wastes are considered as an environmental health hazard.

Potential Environmental and health Impacts from Solid Waste

The typical municipal solid waste stream will contain general wastes (organics and recyclables), special wastes (household hazardous, medical, and industrial waste), and construction and demolition debris. Most adverse environmental impacts from solid waste management are rooted in inadequate or incomplete collection and recovery of recyclable or reusable wastes, as well as co-disposal of hazardous wastes. These impacts are also due to inappropriate siting, design, operation, or maintenance of dumps and landfills. Improper waste management activities can cause:

1. Fire hazards (dry solids / fuel
2. Smell hazard (rotting wastes
3. Fly hazard
4. Air-borne matter hazard
5. Water pollution hazard
6. Aesthetic hazard
7. Blocking of drains / infrastructure e.g. sewers, storm drains and causing flooding
8. Hiding ground for vermin such as rats, snakes
9. Hiding point for criminals
10.   Toilet for passers-by and children

Challenge: Discuss each of these items.

(b) Discuss the concept 'nutrients' as used in liquid waste management

Nutrients: (Phosphates and nitrates)Cause algal blooms with adverse effects on aquatic life.

Phosphate:

Although phosphate is not a toxic substance, excess levels in lake waters can promote eutrophication, the excessive growth of aquatic plants and eventual depletion of oxygen. The major source of phosphate in surface water is from fertilizer. Application practices can cause soil-adsorbed particles to run off into surface water. Over the years, the amount of phosphate used in households is declining. Very few laundry or kitchen products use phosphates anymore. Certain powder dishwashing soaps still contain this substance due to its superior spot resistance. Phosphate is a minor by-product of organic decomposition of sewage, and small amounts of phosphorus are present in sewage. It is almost impossible to link phosphate in the environment with septic systems because the amounts produced are so small when compared with natural sources and surface application of phosphate on farms and lawns.

Nitrates and chemical reactions

Beginning in an enclosed air-tight pool such as a septic tank, organic nitrogen compounds are broken down by mineralization and inorganic ammonium (NH_4^+) is released. Ammonium is soluble in water but is weakly retained in soil by attraction to negatively charged soil surfaces. Under aerobic conditions inorganic ammonium is rapidly oxidized to nitrate (NO_3-) through a microbial process called nitrification. Nitrate as an ion is very soluble in soil solution, and is often leached into the ground water.

The liquid wastes are sometimes called effluent or waste-water. This is typically 80% water, and its quantity is a derivative of per capita water use. In the septic tank, organic nitrogen compounds are broken down by mineralization and inorganic ammonium (NH_4^+) is released. NH_4^+ is soluble in water but is weakly retained in soil by attraction to negatively charged soil surfaces. Under aerobic conditions, NH_4^+ is rapidly oxidized to nitrate (NO_3-) through a microbial process called nitrification. NO_3- is very soluble in soil solution, and is often leached into the ground water (http:/www.economic.com), 2003).

Nitrate poisoning of infants caused the establishment of drinking water standards for this substance. However, NO_3- will continue to be an important indicator of subsurface pollution because it is associated with many other harmful substances that can pollute drinking water. Few municipal systems use any method of nitrogen removal. Municipal sewage treatment, unlike on-site septic systems, concentrates NO_3- at the treatment plant. Typically NO_3- from municipal sewage treatment is discharged underground in huge drain fields or expelled to surface water .

(ii) Total suspended solids as used in water pollution

Comprises large and dense solids or smaller particles some of which can be removed by sedimentation. They cause tirbidity and indicate the presence of pollutants. They may also shield bacteria allowing them to escape the treatment process. Quantities are measured by filtering. They comprise the screenable and non screenable solids. Suspended matter: e.g. clay, silt from soil erosion. Cause turbidity, reduce depth of sunlight penetration, and thus affect photosynthesis. Screenable solids are lumps of material measuring from 25-50mm (1-2 inches) in diameter that can be removed from the water by coarse screening. For example lavatory paper, food particles nappy liners and other non-biodegradable items. • Non-screenable solids are small particles, less than 25mm in diameter, mainly in suspension such as bacteria, faecal particles, fats, oils and soaps.

(c) Write short notes on: (i) physical hazards
(ii) Biological hazards.

Biological Hazards

Biological hazards cause more lost man-hours for inspectors than all the other hazards combined. They consist of micro and macro-biological sources. Microbiological sources include viruses, bacteria and parasites. Every facility is a separate environment where plant personnel bring bacteria and disease into a central location. You should be particularly cautious around food and water sources, rest rooms and washing facilities. Macro-biological sources may cause harm from bites or stings and include things like guard dogs, insects, snakes and other animals. Biohazards also include botanical sources such as poisonous plants and allergic reactions caused by dust or pollen.

Physical Hazards

These include things that cut or crush you, things that you might trip over or fall into or slip on. They also include extremely high or low temperatures, dry or humid atmospheres, poor lighting and excessive noise. The potential of injury from physical hazards may be increased by circumstances where your senses are impaired, such as poor hearing because of hearing protection or an inability to communicate by voice because of excessive noise. Visibility may be impaired from a full-face respirator. Bulky protective clothing may make it difficult to move around in tight spaces. Protective clothing may be a hazard because it is too hot, heavy or bulky. There is a fine line between paranoia and prudent caution but in the end it is always better to be cautious. A thorough and comprehensive understanding of real and potential hazards is best achieved by having a safety conscious attitude.

1)   Explain the concept of oxygen demand (OD) in relation to environmental pollution

OD is the amount of oxygen needed to oxidize a given volume of the organic content of the wastewater-organics and reduced inorganics . It actually refers to the oxidation of the carbon based molecules in the WW, and is used to measure the carbonaceous organic pollution. There is a second stage oxygen demand after about 8 days at 200C, when ammonia is oxidized to nitrates

Theoretical Oxygen demand (OD_{Th})
Is based on chemical reactions, as in oxidation of glucose. $C_6H_{12}O_6 + 6O_2 = 6CO_2 + 6H_2O$ for glucose solution of 250 mg/L,   the theoretical demand is $250 \times 192 / 180 = 266.7$ mg/L
For most sewage, the composition is so complex that the theoretical Oxygen demand cannot be calculated. Thus OD_{Th} has little practical value for WW treatment.

BOD_5 – 5-day oxygen demand
The microorganisms, which break down sewage, are extremely efficient at obtaining and using oxygen. They use up huge amounts of oxygen in the process, suffocating aquatic life and causing the food web to fail. Organic matter uses oxygen when decomposing. Thus a measure of the organic matter in a body of water is given by measuring the amount of oxygen being removed from that water during decomposition. This Measurement is known as the Biochemical Oxygen Demand (BOD). BOD is a measure of the dissolved oxygen consumed during the biochemical oxidation of organic matter present in a substance. standardized tests are done by BOD incubation period lasting five days, at 20^0C (this being the average time taken for sewage to reach the river mouth of an average British river)• This accounts for most, but not all the carbonaceous oxygen demand. . The 5-day period means that nitrification does not start because nitrifying bacteria reproduce more slowly – and usually reproduction starts after 7 days.

Organic compounds:
Organic matter comprises the bulk of the solids in wastewater. Chemical and biological oxygen demand (COD and BOD), total organic carbon, and suspended solids are water quality analyses

commonly used to indicate the amount of organic matter present in wastewater. Nearly all-organic matter in household wastes is biodegradable, and it does degrade readily in soil. When decomposable, they exert an oxygen demand (COD and BOD)

COD:

When the sewage sample contains compounds that are not degradable by microorganisms, or not biologically degraded within the five-day incubation period the Chemical Oxygen Demand (COD) test is used. Chemical oxygen demand (COD) is a measure of the amount of oxygen required to chemically oxidize organic matter in a sample. It is based on laboratory conditions and use acids, rather than bacteria, to break down matter.

COD: BOD = 1: 1.5 – 1: 3 for domestic wastewaters and 1: 3- 1:4 for industrial effluents.

BOD ultimate (BOD_{ult}) is based on breakdown over a longer time – 20 day. Because it includes auto-oxidation phase, it is more representative of both carbonaceous and nitrifying breakdown

Relationship among BOD, COD and TOC

Generally, Od_{th} > COD> BOD_{ult} > BOD_5

The three parameters (BOD, COD and TOC) measure different quantities, and tests take different amounts of time. It is therefore not convenient to use just one of these parameters. For specific wastewater flows it may be possible to monitor each parameter and then establish approximate ratios between the three different parameters, but there is no universal relationship between them. If approximate ratios can be established, then rapid tests can be used for routine monitoring of wastewater characteristics. In some situations it may not be possible to conduct all three tests or to obtain reliable results for all three parameters. Ratios will also depend on the amount of treatment provided to the sample; whether the wastewater is untreated or partially treated. BOD tests are widely used, but the test procedure is lengthy, and results are often of limited accuracy. Oxidation rates are more predictable for dissolved materials than for suspended solids. Opinions vary about which test (COD or TC) provides more reproducible results.

You should only state that 'BOD has no stoichiometric validity' if you understand what this means. (Essentially, the five-day period is arbitrary, and does not correspond to conditions outside the laboratory.)

For a sample of raw domestic wastewater, the value of TOC is likely to be lowest, then BOD, with COD being greatest. Suggested values for the three parameters, using a sample of raw domestic wastewater are:

COD	250 to 1000 mg/l
BOD_5	110 to 400 mg/l
TOC	80 to 290 mg/l

TOC measures carbon, and BOD and COD measure oxygen. TOC is the lowest, because every carbon atom needs two oxygen atoms to be fully oxidised. Carbon has an atomic weight of 12, and oxygen has an atomic weight of 16. Twelve grams of carbon therefore requires about 32 grams of oxygen. Following treatment, TOC may exceed BOD_5.

Consider each parameter in turn:

Chemical Oxygen Demand (COD)

"The Chemical Oxygen Demand (COD) method determines the quantity of oxygen required to oxidize the organic matter in a waste sample, under specific conditions of oxidizing agent, temperature, and time. Since the test utilizes a specific chemical oxidation the result has no definite relationship to the Biochemical Oxygen Demand (BOD) of the waste or to the Total Organic Carbon (TOC) level. The test result should be considered as an independent measurement of organic matter in the sample, rather than as a substitute for the BOD or TOC test."

Chemical oxygen demand (COD) cannot distinguish between organic and inorganic wastes, so the presence of inorganics in a wastewater can increase the COD result. The COD test may be used, however, when the waste contains chemicals that may be harmful to the bacteria used for the BOD test.

Ammonia is not oxidised during the COD test, although nitrite is oxidised to nitrate. Nitrite concentrations are generally very low, however, in comparison to concentrations of organic (carbonaceous) materials.

WATER RELATED DISEASES

Pathogens are disease causing organisms, and are a health hazard. Pathogens are of greater importance in hot climates where survival rates are longer and where people have more contact with water. Our major concern as regards to environmental sanitation is to control diseases related to poor environmental sanitation.

- Water borne diseases – typhoid, cholera, leptospirosis, bacillary dysentery, amoeba, paratyphoid etc. These are transmitted when the pathogens are in water and infect man through drinking water. Chlorination and other treatment is an intervention strategy for this class of diseases.. However, chlorination of sewage is not reliable, because of varying chlorine demands, resistance of some pathogens, production of dangerous chemicals (chlorination byproducts) and subsequent breeding of resistant bacteria

- Water washed diseases e.g. gastroenteritis, ascariasis, relapsing fever, leprosy, scabies etc. Conditions whose occurrence reduce with increased water quantity; i.e. generally result from water scarcity, and reduce regardless of water quality.
- Water based diseases- urinary schistosomiasis and guinea worm diseases; infection due to unfed parasitic worms which depend on aquatic intermediate host to complete their cycle.
- Insect vector borne diseases.

SANITARY INSPECTION
Discuss this under the following subtopics:

       (i) Importance   (ii) when it is done
       (iii) How it is done

(a) Drinking water quality control program involves 2 equally important activities;

- Carrying out the sanitary inspections
- Sampling and analysis of water.

Variations in the quality of water supplies can help in detecting contamination problems, and in determining whether these have arisen at the source, during water treatment, or in the distribution system. However, it may often not be possible to take more than a few samples, and consequently the results of any analysis may not be representative of the water supply system as a while. Microbiol analysis also only gives the results of water quality several days after harm is done;

Water quality (bacteriological) analysis is useful as the sole indicator of the safety of a water supply. Sanitary inspections therefore:

- Allow for an overall appraisal of the many factors associated with a water supply system, including the water works and the distribution system.
- Such an appraisal may later be verified and confirmed by microbial analysis, which indicate the severity of the defects.
- Provide a direct method of pinpointing possible problems and sources of contamination.

- Important in the prevention and control of potentially hazardous conclusions including epidemics of water borne diseases

When to do Inspectors

Routine inspections are visits made with a defined frequency in accordance with a previously established plan. Also non-routine visits by the inspector will be necessary in a typical situations, such as the introduction of a new water source and in cases of emergency.

Emergency situations calling for the urgent present of S.I. include;
- Reports of epidemics
- High morbidity caused by floods
- Unresolved cases where bacteriological analyses repeatedly show the presence of excess levels of microbial organisms and where residual chlorine levels remain the consistently low.
- The detection of any important changes that could impair drinking water quality.

NB: Routine surveillance can be carried out by technical operators so long as they have received suitable training

© Sanitary inspection methodology
- Examine water supply system or at least whether the installations are satisfactory and whether the various operations are being carried out properly.
- Follow the natural sequence, starting from the source of water, its intake, treatment point, disinfection point, storage point and their distribution.
- In each case record what has been observed in appropriate forms, procedures for sanitary inspections should be rapid, systematic and complete at all key points of any water supply system.

Remedial and preventive measures

Surveillance – control the quality of water and protect consumers if sanitary deficiencies identified by such surveillance are not remedial, the effort put into the program has been wasted.
- The reasons for poor water quality should be identified and the cause corrected or eliminated and where necessary emergency precautions taken thus bacteriological, chemical assessment and sanitary inspections give guide to remedial and preventive measures
- Selection of safe and adequate sources

- Constant vigilance – checking of disinfections through residual chlorine tests; commonly education.
- PHC programmes – bacteriological and chemical analysis after remedial measures are taken /implemented.
- Warnings to the community to boil the water or add disinfectant wherever a serious problem occurs – sanitary checks to ensure that the remedial measures have been carried out properly.

Remedial measures
- Cleaning and disinfection of dug wells
- Removal of any cross considerations
- Warnings to the community to boil water
- Disinfection of collected drinking water by the community
- Confirmation that remedial measures have been implemented and that they are effective by means of bacteriological analysis and/or residual chlorine testing.
- Introduction of further sanitary checks preventive measure. Those remedial measures that can be introduced over a period of time are what may be called preventive measures.

WASTE STABILIZATION PONDS AND THE TRICKLING FILTER
Secondary stage: largely a biological treatment (oxidation) phase of settled sewage. It also involves secondary settlement of suspended solids, and uses the PF or AS methods.
- Percolating filter (PF) / the trickling filter /or biological filter)
- Activated sludge (AS) treatment = the aeration of freely suspended flocculants bacteria such that the activated sludge flock in conjunction with the settled sewage which together constitute the mixed liquor.
- Sludge treatment – thickening (by stirring or flotation), digestion aerobically or an aerobically; heat treatment; composting with domestic refuse, chemical conditioning; dewatering on drying beds, vacuum filtration, heat drying; incinerations etc.

Waste water stabilization pond technology is one of the most important natural methods for wastewater treatment. Waste stabilization ponds are mainly shallow man-made basins comprising a single or several series of anaerobic, facultative or maturation ponds. The primary treatment takes place in the anaerobic pond, which is mainly designed for removing suspended solids, and some of the soluble element of organic matter (BOD5). During the secondary stage in the facultative pond most of the

271

remaining BOD5 is removed through the coordinated activity of algae and heterotrophic bacteria. The main function of the tertiary treatment in the maturation pond is the removal of pathogens and nutrients (especially nitrogen). Waste stabilization pond technology is the most cost-effective wastewater treatment technology for the removal of pathogenic micro-organisms. The treatment is achieved through natural disinfection mechanisms. It is particularly well suited for tropical and subtropical countries because the intensity of the sunlight and temperature are key factors for the efficiency of the removal processes. [1].

Waste water treatment in waste stabilization ponds
Anaerobic ponds
These units are the smallest of the series. Commonly they are 2-5 m deep and receive high organic loads equivalent to100 g BOD5/m3 d. These high organic loads produce strict anaerobic conditions (no dissolved oxygen) throughout the pond. In general terms, anaerobic ponds function much like open septic tanks and work extremely well in warm climates. A properly designed anaerobic pond can achieve around 60% BOD5 removal at 20° C. One-day hydraulic retention time is sufficient for wastewater with a BOD5 of up to 300 mg/l and temperatures higher than 20° C. Designers have always been preoccupied by the possible odour they might cause. However, odour problems can be minimised in well designed ponds, if the SO_4^{2-} concentration in wastewater is less than 500 mg/l. The removal of organic matter in anaerobic ponds follows the same mechanisms that take place in any anaerobic reactor. [1]; [2].

Facultative ponds
These ponds are of two types: primary facultative ponds receive raw wastewater, and secondary facultative ponds receive the settled wastewater from the first stage (usually the effluent from anaerobic ponds). Facultative ponds are designed for BOD5 removal on the basis of a low organic surface load to permit the development of an active algal population. This way, algae generate the oxygen needed to remove soluble BOD5. Healthy algae populations give water a dark green colour but occasionally they can turn red or pink due to the presence of purple sulphide-oxidising photosynthetic activity [3]. This ecological change occurs due to a slight overload. Thus, the change of colouring in facultative ponds is a qualitative indicator of an optimally performing removal process. The concentration of algae in an optimally performing facultative pond depends on organic load and temperature, but is usually in the range 500 to 2000 µg chlorophyll

272

per litre. The photosynthetic activity of the algae results in a diurnal variation in the concentration of dissolved oxygen and pH values. Variables such as wind velocity have an important effect on the behaviour of facultative ponds, as they generate the mixing of the pond liquid. As Mara et al. [1] indicate, a good degree of mixing ensures a uniform distribution of BOD5, dissolved oxygen, bacteria and algae, and hence better wastewater stabilization. More technical details on the efficiency of the process and removal mechanisms can be found in Mara et al. [1]; and Curtis [4].

Maturation ponds
These ponds receive the effluent from a facultative pond and its size and number depend on the required bacteriological quality of the final effluent. Maturation ponds are shallow (1.0-1.5 m) and show less vertical stratification, and their entire volume is well oxygenated throughout the day. Their algal population is much more diverse than that of facultative ponds. Thus, the algal diversity increases from pond to pond along the series. The main removal mechanisms especially of pathogens and faecal coliforms are ruled by algal activity in synergy with photo-oxidation. More details on these removal mechanisms in maturation ponds can be found in Curtis.

On the other hand, maturation ponds only achieve a small removal of BOD5, but their contribution to nitrogen and phosphorus removal is more significant.. A total nitrogen removal of 80% in all waste stabilization pond systems, which in this figure corresponds to 95% ammonia removal. It should be emphasised that most ammonia and nitrogen is removed in maturation ponds. However, the total phosphorus removal in WSP systems is low, usually less than 50%

The potential solutions were: activated sludge, aerated lagoons, trickling filter and WSP system. the WSP is considered a most realistic alternative than the rest because of the following: lower energy requirement for pumping; lower Capital cost; less miscellaneous energy costs; less labour costs; less Cost of imported items/ forex; less operation and maintenance costs; high Pathogen removal; superior Protozoa cyst and helminthes removal; General suitability and appropriateness for the hot area; higher Technical feasibility; low Smell hazard with good design. WSPs are cheap to construct and maintain, and have low energy requirements, thus qualifying as the realistic solution. It heavily consumes land, but in a sparsely populated rural town. 2-3 maturation ponds are recommended

THE WATER TREATMENT PROCESS

Simple procedures include:
Physical processes e.g. Screening, degritting, coagulation, flocculation and sedimentation. These are the primary treatment components coagulation and flocculation if accomplished by adding chemicals to modify the pH and enhance the formation of insoluble hydroxides of metals. The latter could be toxic to biological agents and therefore partly help in their removal such chemicals include $NaOH$, Na_2CO_3, $Al_2(SO_4)$ and $Ca(OH)_2$

Then the secondary biochemical treatment is accomplished and micro-organisms are vital here. They are eventually removed by disinfection using chlorine or other chemicals.

You are a medical officer in a district where residents rely heavily on pond water. During a routine assessment of public health in the district, you laboratory technologist informs you that the E. coli count in the majority of the ponds is averagely 1000E.coli / 100ml. According to the Ministry Of Health this 1000 folds exceeds the recommended level for portable water quality. What would you do in this situation?

ASEMBO INTERNATIONAL UNIVERSITY:
SOLID WASTE MANAGEMENT
IMPORTANCE OF SOLID WASTE TO ENVIRONMENTAL HEALTH PROFESSIONALS
Potential Environmental and health Impacts from Solid Waste
The typical municipal solid waste stream will contain general wastes (organics and recyclables), special wastes (household hazardous, medical, and industrial waste), and construction and demolition debris. Most adverse environmental impacts from solid waste management are rooted in inadequate or incomplete collection and recovery of recyclable or reusable wastes, as well as co-disposal of hazardous wastes. These impacts are also due to inappropriate siting, design, operation, or maintenance of dumps and landfills. Improper waste management activities can cause:

1. Fire hazards (dry solids / fuel
2. Smell hazard (rotting wastes
3. Fly hazard
4. Air-borne matter hazard

5. Water pollution hazard
6. Aesthetic hazard
7. Blocking of drains / infrastructure e.g. sewers, storm drains and causing flooding
8. Hiding ground for vermin such as rats, snakes
9. Hiding point for criminals
10. Toilet for passers-by and children

(i)Increase disease transmission or otherwise threaten public health. Rotting organic materials pose great public health risks, including, as mentioned above, serving as breeding grounds for disease vectors. Waste handlers and waste pickers are especially vulnerable and may also become vectors, contracting and transmitting diseases when human or animal excreta or medical wastes are in the waste stream. (See the discussion on medical wastes below and the separate section on "Healthcare Waste: Generation, Handling, Treatment, and Disposal" in this volume.) Risks of poisoning, cancer, birth defects, and other ailments are also high.

(ii) Contaminate ground and surface water. Municipal solid waste streams can bleed toxic materials and pathogenic organisms into the water bodies. (Leach ate is the liquid discharge of dumps and landfills; it is composed of rotted organic waste, liquid wastes, infiltrated rainwater and extracts of soluble material.) If the landfill is unlined, this runoff can contaminate ground or surface water, depending on the drainage system and the composition of the underlying soils.

Many toxic materials, once placed in the general solid waste stream, can be treated or removed only with expensive advanced technologies. Currently, these are generally not feasible in Africa. Even after organic and biological elements are treated, the final product remains harmful.

(iii) Create greenhouse gas emissions and other air pollutants. When organic wastes are disposed of in deep dumps or landfills, they undergo anaerobic degradation and become significant sources of methane, a gas with 21 times the effect of carbon dioxide in trapping heat in the atmosphere.

Garbage is often burned in residential areas and in landfills to reduce volume and uncover metals. Burning creates thick smoke that contains carbon monoxide, soot and nitrogen oxide, all of which are hazardous to human health and degrade urban air quality. Combustion of polyvinyl chlorides (PVCs) generates highly carcinogenic dioxins.

(iv) Damage ecosystems. When solid waste is dumped into rivers or streams it can alter aquatic habitats and harm native plants and animals. The high nutrient content in organic wastes can deplete dissolved oxygen in water bodies, denying oxygen to fish and other aquatic life form. Solids can cause sedimentation and change stream flow and bottom habitat. Siting dumps or landfills in sensitive ecosystems may destroy or significantly damage these valuable natural resources and the services they provide.

(v) Injure people and property. In locations where shantytowns or slums exist near open dumps or near badly designed or operated landfills, landslides or fires can destroy homes and injure or kill residents. The accumulation of waste along streets may present physical hazards, clog drains and cause localized flooding.

(vi) Discourages tourism and other business. The unpleasant odor and unattractive appearance of piles of uncollected solid waste along streets and in fields, forests and other natural areas, can discourage tourism and the establishment and/or maintenance of businesses.

CHARACTERISATION OF SOLID WASTES
(b) (i). What is characterization of solid waste
(ii). Explain the significance of each aspect

The following characteristics of waste comprise waste characterization: waste density, composition, moisture content, size distribution generation rate and the reasons for their choice is also discussed under each item. An effective SWM plan requires full knowledge about the nature of the wastes. This can be determined by sampling. Solid wastes can then be characterized using the following criteria: Generation rate (ii) Moisture Content (iii)_ Density (iv) Composition (iv) Size Distribution

After assessing the quantity, the other characteristics of waste to enable one select the most appropriate waste disposal option include: Moisture content, biodegradability potential, heating value and density, composition, and size distribution. An effective Solid Waste Management (SWM) requires these details because they will affect choices concerning:

(i) Method of storage (ii) The method and frequency of collection; (iii) The equipment used for collection; (iv) The size of the workforce; and (v) The method of disposal. In addition, the properties of the waste also indicate: The potential for resource recovery; and (ii) the environmental impact if the wastes are not properly managed.

Waste generation Rate:

This aspect describes how quickly a certain quantity of waste is produced. . It is usually defined as the average amount of waste generated by one person in one day. In weight terms, the common unit is Kilograms per capita per day (Kg/cap/day) (hereafter called Kg.c.d.). Thousands of tons of solid waste are generated daily in Africa. Most of it ends up in open dumps and wetlands, contaminating surface and ground water and posing major health hazards. Generation rates, available only for select cities and regions, are approximately 0.5 kilograms per person per day—in some cases reaching as high as 0.8 kilograms per person per day. While this may seem modest compared to the1–2 kg per person per day generated in developed countries, most waste in Africa is not collected by municipal collection systems because of poor management, fiscal irresponsibility or malfeasance, equipment failure, or inadequate waste management budgets.

Waste composition:

The composition of the seven solid waste types is highly variable. They are influenced strongly by: climate and seasonal variation, the prevailing economy, the physical characteristics of the city, social and religious customs. Variations in solid waste composition are mostly significant when a municipality is making decisions about the suitability of a specific treatment or disposal method, such as composting or incineration. It is important for the waste manager in a city to have an indication of the composition of waste in the locality. Sampling of the wastes is technically possible, and must be done in one or more areas, which are statistically representative. In this way, the information on waste composition and characteristics can be obtained to aid in decision-making about the viability of waste treatment processes, collection equipment changes and recycling initiatives. Flintoff, 1976, cited in Ali (2003), developed the following model of changes in waste quantities at different stages in the waste handling processes.

Figure: Changes in composition and nature of solid waste along the stream (Source: Ali, 2003)

Stage		HANDLING PHASE	LOSSES
TOTAL GENERATED			
1	Minus	Salvage sold by householder	
2	Minus	Salvage by servants	
3	Minus	Salvage by waste pickers in streets	
= TOTAL COLLECTED			
Minus		Salvage by collectors	
=TOTAL DELIVERED FOR DISPOSAL			
Minus		Salvage by disposal staff	
Minus		Salvage by waste pickers at disposal site	
= TOTAL DISPOSED			

Moisture Content

This is the percentage by weight of waste that is water. Drying a known amount of waste and measuring the weight change can determine this. Moisture Content (MC) is a measure of the amount of water in the waste. It is usually expressed as a percentage of the weight of water in a substance compared to its total weight. It is measured by noting the loss of weight after drying. MC depend on the presence of certain components, and is often related to the proportion of food wastes in a sample, as this component often has approximately 70% MC. MC is a vital factor in the choice of final disposal of waste. For instance, wastes with high MC cannot be incinerated; composting requires a certain amount of moisture; while sanitary land filling depends on bio-degradability of waste, which is affected by MC. The level of moisture in a waste also determines the generation of polluted water (Leachate) and breeding of flies.

Waste density

The density of an object is the mass of an object made of that material divided by its volume. It is normally measured in Kilograms per m^3 (Kg/m^3). For solid waste, the term bulk density should be used since there is a considerable amount of air between the individual pieces of waste material. Bulk density includes the volume of this interstitial air. The bulk density of waste increases as the waste is compresses, or if water is added to it (Ali, 2003). Density of waste can be used in conjunction with generation rates to estimate the volume generated. This can vary from 0.4-1.0 liters / person/day (lcd). The volume and density of waste are important in choosing both the size and type of containers and collection vehicles. For instance, if a container of a certain capacity is chosen, the density can be used to calculate its weight when it is full to indicate whether it will be possible for one man to lift the full container.

The average density of solid wastes tends to decrease with level of development. In developing countries, there are lots of bulky waste material such as organic matter, mud, ash and soil. Besides, there is a lot of scavenging as some material is salvaged for sale, reuse or recycling. The ash volume increases because of widespread use of solid fuel.

This is important in each stage in collection and disposal. For planning waste collection, the density achieved in each collection vehicle influences how many are needed to collect the waste in a particular area. Waste density is also important in a landfill site as an indication of how much space each delivery of waste will take before and after compaction and as a consequence, how long the total landfill space will take.

Size Distribution

Waste comprises many separate objects of different sizes. The size may influence the collection and disposal methods used, such as the diameter of storage bins. Large wastes may sometimes require size reduction by breaking up through shredding. This may facilitate compressing at the landfill, decomposition during composting and / or packaging in collection and storage bins. This may lead to a more uniform waste that may be easier to handle and process. Wastes with fine constituents such as sand may require screening prior to any further processing, as sand and ash can cause serious abrasion if the waste is made to slide against parts of a vehicle body. This may lead to quicker depreciation of the vehicle.

COMPOSTING OF SOLID WASTES

What is composting as used in solid waste management?
Meaning of Compost:
Compost is the humus-rich, decomposed organic matter. It may be produced by aerobic or anaerobic process. Aerobics composting is more popular and common because the final product has no objectionable smell.

Methods of composting

Among the most common composting method is the windrow composting Ali, 2003). This involves arranging the organic waste in rows and either aerating by passing air or blowing used air from it. This replenishes the oxygen supply to the decomposing matter. Regular turning of the waste is necessary to ensure that all waste get a chance of being at the centre whose temperature of 700C kills most weed seeds, pathogen eggs, parasites and other undesirable living organisms in the solid waste.

Vermicomposting

Worms can be introduced into the organic matter to decompose it faster. The worms eat up the organic matter and throw a solid waste called castings. This is rich in many minerals that plants require for normal growth. Decomposed organic matter (hereafter called compost), is rich in a number of macro and micronutrients.

Importance of compost in Agriculture
Compost is mostly used in farms as soil conditioners. It adds both macro and micro nutrients to the soil; it provides food for soil microbes- thereby making the soil healthy; it helps stabilize and build up the soil structure (mostly produces a crumb structure); it improves soil moisture and nutrient holding capacity; it improves drainage and aeration of the soil; and it cushions soils against extreme temperature changes (since it is spongy and therefore has air which is a bad conductor of heat). Therefore, it is vital that any compost produced should be able to perform the above roles effectively. This is why marketing and understanding of demand come in.

LANDFILLS

Describe the different kinds of landfill.
There are three types of landfills according to the state of California (1984) (Tchobanoglous et al 1993). These are:
(i)   Conventional landfills for commingled MSW
(ii) Landfills for milled solid wastes, and
(iii)   Monofill for designated or specialized wastes
 iv.   Others e.g. (a) maximum gas production system; (b) integrated solid waste treatment units (c) wetland landfills

There are three main types of landfills These are: (i) Conventional landfills for commingled MSW (ii) Landfills for milled solid wastes, (iii) Monofill for designated or specialized wastes (iv) Others e.g. (a) maximum gas production system; (b) integrated solid waste treatment units (c) wetland landfills. In places without appropriate cover material, e.g. where all soil is sand, it would require that appropriate material to be imported from elsewhere. To reduce this expense, it would be better to consider seriously the integrated solid waste treatment units. This involves the organic constituents being separated out and placed in a separate landfill where the biodegradation rates would be enhanced by increasing the moisture content of the waste, either by recycling leachates or by seeding with digested wastewater treatment plant sludge or animal manure. The degraded material would be excavated and used as cover material for new fill areas, and the excavated cell would be filled with the new waste. This can give additional landfill capacity.

Sanitary landfilling is a fully engineered disposal option. It avoids the harmful effects of uncontrolled dumping by spreading, compacting and covering the waste on land that has been carefully engineered before use. Through careful site selection, preparation and management, operations can minimize risks from leachate and gas production both in the present and the future. Site plans and design consider not only waste disposal but aftercare and ultimate land use once the site closes. Sanitary landfill is suitable when suitable land is available at an affordable price. This option must be considered after an assurance that pollution could be controlled and human and technical resources are available to operate and manage the site. Ali et al. (1999) state that hazards arising from landfill can vary from one site to another, but depend primarily on a range of factors including waste composition, moisture and climate.

Sanitary landfills.
Sanitary landfills are the only land disposal option that enables control and effective mitigation of Potential surface and groundwater contamination; Health and physical threats to waste pickers and sanitation workers; and methane emissions. Sanitary landfills require much greater initial investment and have higher operating costs than controlled dumps. Full community involvement throughout the life cycle of the project is essential. Proper design, operation and closure also require a much higher level of technical capacity.

Siting.
Siting is possibly the most difficult stage in landfill development.
1. Carry out an environmental impact assessment that addresses all siting criteria (see box at left).
2. Organize full community involvement. This is especially important given the greater expense and often greater size of sanitary landfills.

Design.
To mitigate environmental impacts, sanitary landfill designs should include:
1. An impermeable or low-permeability lining (compacted clay and polyethylene are most common in developing countries; geopolymers and asphalt are prevalent in the developed world).
2. Leachate collection, monitoring, and treatment.

3. Gas monitoring, extraction, and treatment.
4. Fencing to control access.
5. Provisions for closure and post-closure monitoring and maintenance.

Leachate management.
Leachate impacts can be controlled only with lined landfills.
1. Install collection systems to retrieve leachate from the bottom of the landfill.
2. Treat leachate physically, chemically, or biologically through:
a. An off-site sewage treatment plant (adequate sewage treatment facilities are readily available in only some parts of Africa), or in a dedicated on-site treatment plant.
b. Recirculation that sprays leachate from the bottom of the landfill onto its surface. This is a popular landfill management practice in Africa. It reduces leachate volume by increasing evaporation, stores remaining leachate in the body of the landfill, and may accelerate degradation and extend the life of the site. However, recirculation is a new technique whose long-term effects are not yet known.
c. Evaporation of leachate through a series of open ponds. This method requires pumping and some means for disposing of possibly toxic residues. Ponds should be designed with enough capacity to accommodate increased volume during the rainy season.
3. Monitor groundwater and surface water regularly, both down-gradient and up-gradient from the landfill. At a minimum, monitoring should include indicators of core contaminants, chemical oxygen demand, biological oxygen demand, and total nitrogen and chloride levels.
4. If it is uneconomical to recover and use landfill gas as fuel, it should be vented and flared. Currently, recovery and processing systems are both expensive and difficult to operate. These systems are economical only when the landfill generates large quantities of gas, where local or regional demand exists, or where the price for natural gas or other substitutes is high. At a minimum, buried perforated pipes that can safely vent gas should be installed, and a flaring system should be added to reduce global methane release to the atmosphere.
5. Fence in landfills to prevent waste pickers from accessing the site. This enables landfill personnel to work efficiently and protects waste pickers from exposure to harmful substances. However, it also deprives them of their livelihood. They should thus be integrated into formal collection or disposal operations by, for

instance, helping them organize a cooperative and offering them structured access at the landfill gates. Also, they should be made a part of the earlier stages of the collection process, perhaps by helping them establish a cooperative that collects recyclables from industry.

6. When the landfill is full, implement the activities specified in closure and post-closure plans that were developed during design. These should include sealing the landfill and applying a final cover (including vegetation) to it, land use restrictions on the old landfill and surrounding areas, and long-term gas, leachate, surface water and groundwater monitoring.

Describe the impacts of a fully engineered solid waste management system clearly indicating
     (a) The pollution concerns and how they are sorted out.
     (b) The site selection and
     (c) Site operation

Control of landfill gases

Landfill gases must be controlled for as long as they are expected to be generated after the landfill is closed. This may be by use of extraction wells, collector and transmission piping, and gas flaring and/ or combustion facilities. A means of monitoring, collection and management (e.g. by flaring, or re-use) of landfill gas (LFG) should be in the landfill design right at the conception stage. The LFG management should be done throughout the design, construction, use, closure and post-closure stages to avoid fire and air pollution hazards.

Collection and treatment of Leachate

Landfill Leachates are liquids washed from the landfill wastes. They largely comprise organic acids, and have high Biochemical Oxygen demand (BOD) and Chemical oxygen Demand (COD). Leachates therefore have the capacity to contaminate groundwater, but are also able to transport dissolved organic substances that may be released in the unsaturated subsurface environment, by the change in the partial pressure of the constituents in the gas phase. There should be a proper Leachate collection and monitoring strategy throughout the entire life of the landfill (i.e. from conception to post-closure). Their characteristics change with age of the landfill, and these changes should be monitored to be sure they are steady. Any major deviation from normal pattern should be studied and monitored more closely, as it may indicate an interaction with other media such as ground water. This is best done by environmental monitoring systems.

SLF Location
- Careful siting to minimize groundwater and other potential pollution problems;
- Ideally sited away from present and proposed residential areas but not to the extent that transport costs become unaffordable;
- Adequate barriers to protect nearby residents where present;
- Control of wind blown litter (paper, plastics etc) by screening and cover.

Siting guidelines for landfills
Do not site landfills:
- In wetlands or areas with a high water table
- In floodplains
- Near drinking water supplies
- Along geological faults or seismically active regions
- Within two kilometers of an airport

Do site landfills:
- Above clay soils or igneous rock
- With active public involvement
- In areas with sufficient capacity

SLF Operation
a. Minimize contact between waste and water;
b. Compaction of refuse and covering to prevent nuisance through flies and vermin;
c. Prevent the formation of pools of water where mosquitoes could breed;
d. Discourage rodents and enable early discovery of burrows through monitoring;
e. Minimize smells and prevent burning by compacting and covering refuse and controlling site operations;
f. Fill depressions so that profile is uniform;
g. Control birds through prompt covering of waste.

SLF Management and control
- Make site ownership and responsibility clearly identified;
- Earmark site officially and ensure it as actually used by persons allowed at the site;
- Actively monitor and control site operations;
- Restore the site to an acceptable condition after closure;

- Plan for future use of the site.

RECOVERY AND SEPARATION OF SOLID WASTES
(i) Separation at source.
(ii) Material recovery facility

Low participation of households.
Households may not participate in waste management programs because they may view solid waste management ......a low priority. They may be they are unwilling to participate in collection systems or in keeping public spaces clean, or they are unwilling to pay for service. Community Provisions for education, is often key to overcoming the best counter to these barriers, may be inadequate in but traditional approaches to waste management often do not provide enough for education. Community-based solutions can use preliminary research and input from the community to generate a list of desired services, appropriate incentives for households and servants, and systems for cleaning streets and other public places.

Facilitate separation at disposal site.
When waste pickers are allowed access to disposal sites, significant amounts of material can be recovered. However, because they interfere with efficient operation of dumps and landfills, waste pickers are usually excluded from these sites, lowering recovery rates and causing severe economic hardship. Some sites provide a measure of structured access to waste pickers—at the Bisasar Road landfill in Durban, for instance, registered pickers from an adjacent squatter settlement are allowed into the site after hours, earning US$77 per month from this activity. At all other times, armed guards restrict access to the site. Similarly, the South African Boipatong landfill limits access to 100 registered waste pickers.

Management problems.
Problems with traditional waste management schemes include ineffective, inefficient, or unrepresentative management, as well as lack of community accountability to the community. CBM can introduce performance control techniques, share management with an NGO, adjust or by-pass an existing management committee, and provide incentives for managers, such as training and exchange visits.
Operational problems.
With poor motivation operators are poorly motivated, due to low salaries, low status and bad working conditions, operator

motivation can be low, and public service may become can often be unreliable. Finding adequate space for waste facilities and equipment is another potential operational issue. Sound CBM can addresses motivational problems by involving operators in decision-making, using special group incentives, and, in some cases, by granting exemptions from municipal taxes. Operators can be officially introduced to households and provided with identity cards to improve operator status. Space problems can be resolved by lobbying municipalities and local leaders, as well as conducting media campaigns in the neighborhood.

Financial difficulties.
Public and private management plans often face financial difficulties caused by inadequate fee collection and inability to pay for service in low-income neighborhoods. CBM gives community input into plans for fee collection payments, incentives and sanctions for non-payment. Community input can also help waste management providers find lead to additional revenue- generating services.

Lack of municipal cooperation.
If waste collection between the municipal government and private operators is badly coordinated and the community may lose interest in trying to improve the waste situation. Extending service, mobilizing communities to lobby the municipality for assistance, involving local authorities, and structuring formal and informal opportunities for cooperation all improve municipal performance and community support for waste management plans and programs.

Capacity Building
Insufficient capacity is a fundamental impediment to sound solid waste management programs in much of the developing world. Operating an efficient, effective, environmentally sound municipal solid waste management program requires building administrative capacity for government and private sector players and technical capacity for designing, operating, maintaining, and monitoring each part of the process. Often those people working in solid waste management—private sector companies, NGOs, and government entities—lack the technical and financial knowledge to operate efficiently. Training that builds human resource and institutional capacity at appropriate levels is essential. Peer-to-peer training for everyone from waste-picker to local government officials has proven effective in extending and sustaining these programs.

Current Recycling programmes and other available recycling opportunities

Recycling is the primary use of a waste (in its intact form) or secondary use of whole or part of the waste after minor or major modifications (Ali, 2003). Recycling reduces the quantity of waste to be disposed of in a landfill, and is one of the most environmentally friendly waste management methods. The re-used items would gradually reduce in volume through wear and tear (e.g. old clothes used ad duster, mattresses, brooms etc.

Reduce, reuse, recycle (RRR).

Reducing the quantity of waste that must be transported and disposed of should be a primary goal of all municipal solid waste management programs. Waste should be recovered at the source, during transport or at the disposal site. The earlier the separation, the cleaner the material, and, in the end, the higher its quality and its value to users. Incentives which integrate and foster the involvement of the informal sector—itinerant collectors, microenterprises, cooperatives—can be essential to improved waste minimization. Other tips on reducing waste include: Encouraging recycling can help build capacity among local micro-enterprises and reduce the waste handled by landfills and dumps.

HAZARDOUS WASTES (HEALTH CARE WASTES)

Identify the main source and components of health care wastes.

HCW is the total waste stream generated nursing homes, hospitals, mobile surgeries, dental surgeries, health care establishments, research facilities and laboratories. Other names for this kind of waste are clinical waste, Medical waste, and hospital waste.

This includes wastes from health posts, clinics, hospitals, and other medical facilities pose serious and urgent problems in the Africa region. These wastes can contain highly infectious organisms, sharp objects, hazardous pharmaceuticals and chemicals, and even radioactive materials. Since the various forms of healthcare waste require different types of treatment, they should be segregated at the source. General waste should be segregated from hazardous material to reduce volume: sharps should be placed in puncture-proof containers, infectious waste separated for sterilization, and hazardous chemicals and pharmaceuticals segregated into separate bins. Unfortunately, all of the available disposal options are imperfect. The most immediate threat comes from highly infectious waste. On-site treatment is generally preferred to reduce the risk of disease transmission to waste handlers, waste pickers and others. Suggested mitigation measures include:

• In rural areas, burn infectious waste in a single-chamber incinerator, if possible. This kills >99 percent of the organisms and is the best option for minimal facilities.

• In urban areas, burning is not advisable, as the fly ash, toxic gases and acidic gases pose a much greater health threat in more densely populated urban environments than in rural areas. Thus larger facilities should autoclave infectious waste. While high-temperature incineration is theoretically the best option in urban environments, in practice the equipment is rarely operated properly and disposal is highly polluting.

• In some large cities, off-site wet thermal, microwave or chemical treatment options may be available.
• The least expensive option is land disposal. If waste is to be disposed of in a dump or landfill, it should be packaged to minimize exposure, placed in a hollow dug below the working face of the landfill, and immediately covered with 2 m of mature landfill waste. Alternatively, it may be placed in a 2 m deep pit and covered in the same manner. Waste picking must then be prevented.

HCW is the total waste stream generated by nursing homes, hospitals, mobile surgeries, dental surgeries, health care establishments, research facilities and laboratories. Other names are clinical waste, Medical waste, and hospital waste. Ali (2003), states that 75-90% of health care waste can be classified as non-clinical or general waste, which present no higher risk to the community than municipal waste. The remaining portion of 10-25% can be classified as clinical waste, and it is this portion that can be hazardous.
This classification is further illustrated by the figure below.

Constituents of health care waste:

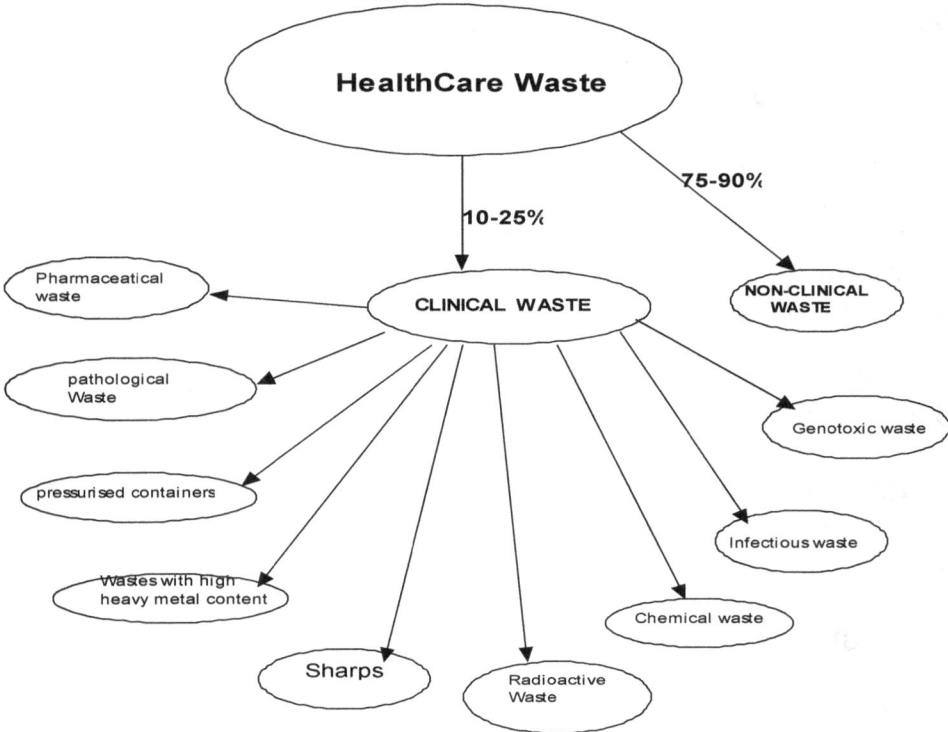

THE CATEGORIES OF HEALTHCARE WASTE (SOURCE: ALI 2003 pp 11.3)
Cited in Afullo (2004)

RISKS FROM HEALTH CARE WASTES
Explain the health risks associated with health care wastes in (a) above

The following categories of HCW exist:

1. General Waste (Non-toxic) e.g. food remains, containers
2. Pathological waste: Solid body parts with cancerous growth e.g. breasts
3. Pharmaceutical waste: Dirty or expired drugs
4. Radioactive waste (from radiology and X-ray laboratory)
5. Wastes rich in heavy metals e.g. Thermometers (mercury), Lead acid batteries (Lead);
6. Combustible waste e.g. cotton wools impregnated with methylated spirits
7. Explosive-prone wastes e.g. old unused gas cylinders;
8. Sharps e.g. needles, scalpel blades, broken drug bottles;
9. Infective waste e.g. blood-stained clothes and equipment;

SOME QUESTIONS IN LIQUID WASTE MANAGEMENT

1. (a) Explain the importance of treating effluents

Essence of WW treatment: To (i) reduce the spread of diseases by removing the majority of pathogens; (ii) reduce the concs of life threatening chemicals to acceptable levels; and (iii) reduce environmental damage from polluting matter present in WWs. Treatment must aim at some standards.

Composition of sewage: this relates to hydraulic and organic loadings, among other parameters.

Parameters of traditional importance
TSS,   Oxygen demand (COD, BOD etc),   pathogens (E coli), Toxic chemicals.

Those of growing importance:
Nutrients,   Helminth eggs and   colour/pH/ammonia.
Dissolved material
may contribute to the colour of the water but not its cloudiness. These consist of organic matter such as proteins, carbohydrates and fatty acids (read COD / BOD) and inorganic ions such as ammonium, chloride, nitrate, nitrite, phosphate, toxic chemicals etc.

Toxic chemicals:
Heavy metals in industrial areas, phenols and various chemicals (e.g. surfactants) may be toxic to human life domestic animals, wildlife and aquatic organisms.

Nutrients: (Phosphates and nitrates) Cause algal blooms with adverse effects on aquatic life.

Colour, Ph and ammonia:
Colour is of aesthetic importance. , While the pH and ammonia levels may have an effect on aquatic life.

Ammonia:
Ammonia is the next most important substance to remove from sewage. Domestic waste can produce large amounts of ammonia, a substance to which many aquatic organisms are highly sensitive.

TSS:
Another easily measured quantity is the amount of matter present in the water. This is known as the Total Suspended Solids (TSS). Thus BOD and TSS are the two sewage qualities most commonly measured by the authorities. Contributes to colour.

Distinguish between DWF and dwf

Dry weather flow: applies to combined sewerage systems, but can also be applied to flows in rivers and streams.

The dwf will vary during the day (because of variation in amount of WW), and could also vary between seasons (e.g. increase in infiltration or changes in population during holidays).

DWF (in uppercase) is the average value, sometimes averaged over a whole year. The average flow during a day is the dwf (in small letters). DWF and dwf therefore differ with respect to time/ duration of coverage. Dry weather flow consists of domestic and industrial wastewater, but not storm water.

Explain: Combined sewers, Separate sewers;, Partially separate sewers
2 marks for each item
(i)   Combined-taking grey water, black water storm water and infiltration inflow;
(ii)   Separate: one system taking grey water and black, and another talking storm water and infiltration / inflow;
(iii)   Partially separate: taking grey water, black water and in some parts of the system, storm water and inflow.

Explain the effluent treatment regime in relation to: 3xDWF and 6 x DWF
i.   Flows of upto 3xDWF = Full treatment to all flows;
ii. From 3xDWF up to 6xDWF gets full treatment provided to a portion of the flow upto 3xDWF; screening, grit settlement and storm water settlement provided for the remainder of the3 flow; until storm tanks are filled after which additional flows are discharged without storm water treatment.
iii. Flows in excess of 6xDWF = full treatment provided to a portion of the flow (3xDWF); screening, grit settlement, and storm water settlement provided for another portion of the flow (also 3xDWF), until storm tanks are filled after which additional flows   are

discharged without storm water settlement. Screening provided ro4r the remainder of the flow in excess of 6X DWF.

Distinguish between BOD5 and BODult

BOD_5 – 5-day oxygen demand
The microorganisms, which break down sewage, are extremely efficient at obtaining and using oxygen. They use up huge amounts of oxygen in the process, suffocating aquatic life and causing the food web to fail. Organic matter uses oxygen when decomposing. Thus a measure of the organic matter in a body of water is given by measuring the amount of oxygen being removed from that water during decomposition. This Measurement is known as the Biochemical Oxygen Demand (BOD). BOD is a measure of the dissolved oxygen consumed during the biochemical oxidation of organic matter present in a substance. standardized tests are done by BOD incubation period lasting five days, at 20^0C (this being the average time taken for sewage to reach the river mouth of an average British river)• This accounts for most, but not all the carbonaceous oxygen demand. . The 5-day period means that nitrification does not start because nitrifying bacteria reproduce more slowly – and usually reproduction starts after 7 days.

BOD ultimate (BOD_{ult}) is based on breakdown over a longer time – 20 day. Because it includes auto-oxidation phase, it is more representative of both carbonaceous and nitrifying breakdown

An industry releases effluents with the following characteristics:
BOD5 = 500mg/L
Temp = 60°C
Discharge = 100L/s == 360m³/hr

The effluent is to be discharged into a river with the following characteristics:
BOD_5 = 2mg/L;
Temp = 20°C;
Discharge flow = 3000m³/hr

MASS BALANCE
Using the mass balance concept and equation; $F1Ci + F_2C_2 = (F_1+F_2)C3$
Issue 4 marks for correct conversions into harmonized units; eg 100L/s=360m³/hr or 3000m³/hr= 3L/hr = 72 L/day

Correct insertions into the mass balance equation for each parameter- issue 3 marks each;

Correct answers – I mark each.

Therefore $C_3 = F_1C_i + F_2C_{2/}C_3$

BOD_5 (3) = 360m³/hrx500mg/L + 3000m³/hr x2mg/L/360 + 3000 m³/L =

180,000 + 6,000 / 3360 = 186,000 / 3360 = 55.357 mg/L

Therefore the BOD_5 at the point of entry of effluent to the river = 55.357 mg/L

Temperature at entry point = 360m³/hrx60°C+ 3000m³/hr x20°C/360 + 3000 °C = 21600 + 60,000 / 3360 =24.286°C.

Therefore the temperature of the water-effluent mixture at the point of entry into the river = 24.3°C

STAGES IN WASTE WATER TREATMENT

Use a well-labeled diagram to explain the preliminary and primary wastewater treatment stages.

• Preliminary treatment – involves removal of easily separated solids; involve screening, whose solid products are then burnt, incinerated or buried or grit – dumped on site. The process helps remove larger floating and suspended materials, render the sewage more amenable to subsequent treatment by removing large and coarse objects which could form blockages, withhold chemicals, act as hiding points for micro-organisms, or damage equipment. Other steps include grit and grease removal and possibly flow measurements.

• Primary stage involves settlement of some suspended solids, sedimentation, removal of scum and sludge. The removal of particles during this stage is controlled by the settling characteristics of the particles e.g. Density, size, ability to flocculate, retention time in the tank and surface washing

Illustrations:

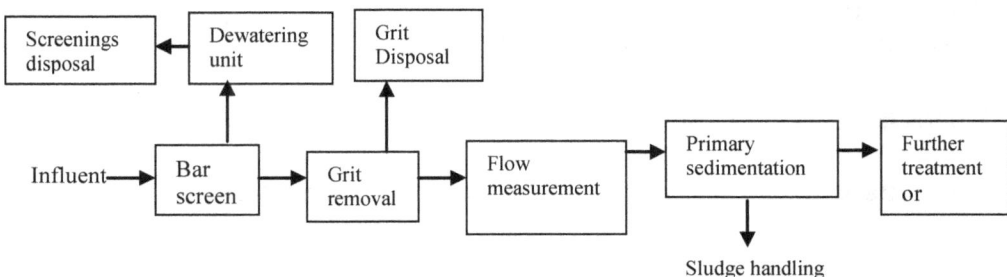

SETTLING AND STROKE'S LAW

Describe the settling behaviour in a sedimentation tank

Types of settling;
- Class I settling – as per stroke's law
- Class II settling – velocity aspect
- Class III settling – zone settling or hindered settlement
- Class IV settling – compressive settling

Settlement conditions are different. Conditions in a primary sedimentation tank are very calm, whereas some turbulence in grit removal units helps to maintain light organic solids in suspension.

Particles in a grit channel settle discretely as individual dense solids. Particles settle with a fairly constant velocity. Particles in a sedimentation tank are light solids that may coalesce (flocculate). Particles combine and their settling velocity increases as they grow larger. Some sedimentation in sedimentation tanks may be hindered or compressive. Settlement of grit usually obeys Stokes' Law. Settlement in a sedimentation tank may be empirical, because wastewaters compositions may vary.

Grit removal principles:
The principle is to reduce the WW flow velocity so that the grit settles, but kept high enough to retain OM in suspension. The theory of grit settlement is based on stroke's law, which provides the maximim settling vewlocity (Vmax) for spherical particle ina liquid.

$$Vmax = {}^2/_9 \ (r^2 g/\eta) \ \rho - \rho'$$

Vmax = velocity of settling particle (m/s);
g= gravitational constant (N/kg or m/s^2);
R= radius of settling particle (m);
P = density of particle (kg/m^3)
P' = density of liquid (kg/m^3)
η = dynamic viscosity of liquid (N.sec/m^2)

SEDIMENTATION TANKS IN WW TREATMENT

Describe the stages of activities in a sedimentation tank

The aim of primary treatment is to provide period during which the WW is stored under calm conditions. The conditions encourage many light solids to sink to the base of the storage tank as sludge and floating material to rise as scum. Primary treatment is a physical treatment stage. Sludge and scum can be separated and removed from the WW during primary treatment, reducing the loading for biological treatment that follows.

The 4 stages are: (i) Inlet zone (ii) Settlement zone (iii) Outlet zone and (iv) Sludge collection and storage zone zone.

Inlet zone: this is where the flow energy of the incoming liquid is dissipated. Should take up only a small section, and cause minimum, if any turbulence. If improperly designed, the flow passing into the settlement zo9ne may still be turbulent.

Settlement zone: represents the true tank capacity, where settlement is accomplished;

Outlet weir: collects the settled WW. The weirs are sometimes V-notched and must always be protected by a scum board-particularly in and partially out of the water- on the tank side o0f the weir, to prevent the loss over the weir of any floating material

Zone of collection and storage of sludge and from where the sludge will periodically be withdrawn either by pumping or under hydrostatic head.

WASTE STABILIZATION POND DESIGN

An urban centre has a population of 8,000 and records an annual population growth rate of 2%. Making assumptions where necessary, design an appropriate waste stabilization pond.

Assumptions:
Design life = 25 years;
Water demand level per capita in 25 years' time = 100 lcd.
% of used water constituting effluent = 80%
Design population corresponds to the population in 25 years time.
Thus:
Design population = $8,000(1+ 0.02)^{25}$ = 8,000 x 1.641 = 13,124.847955717839294150752601141 = 13,125.
Calculate the total daily wastewater flow (Q m^3/d)
Q = per capita WW flow per day X Estimated future population = = 13,125 x 100 x 80%litres = 1050000l/day = 1050m^3/day

Assumptions and background calculations
Design aspects for each of Anaerobic, facultative and maturation ponds

Rationale:

At WSP systems serving more than 10,000 people, it is sensible to have 2 or more ponds in a parallel. The series are equal, and receive the same flow. Design should allow for operation and maintenance when some gates may have to be closed e.g. during desludging, repair (e.g. leakage sealing) etc. It also caters for earlier commissioning of facultative ponds before the rest.

Basic design principles

i. Design with the lowest temperature since the performance of anaerobic ponds increase significantly with temperature. Thus a design temperature has been $22^{0}C$.

ii. Since rectangular shapes give maximum path of flow for the effluent (diagonal), thus reducing / minimizing hydraulic short-circuiting, a rectangular shape has been adopted in the design of all ponds.

iii. Vehicular access to all ponds should be possible in all seasons; a distance of 5m is left between any 2 ponds to facilitate this.

iv. If wind direction is seasonally variable (and it indeed is, as shown above), then wind direction in the hot season should be used as this is when thermal stratification is at its greatest. January is the hottest month, thus the then wind direction is used in the design.

v. To facilitate wind-induced mixing in the surface layers, locate the ponds such that the longest dimension (diagonal) lies in the direction of the prevailing.

vi. To minimize hydraulic short-circuiting, inlet should be located such that the waste-water flows in the pond against the wind direction..

vii. Each pond should have at least 100, preferably more than 200 m across its longest dimension against the wind. Since the maturation ponds have lengths in excess of either of these minima, the angle against the wind has not been given priority.

THE DESIGN:

Choose Suitable slopes for the pond embankments, and a suitable freeboard.

Embankment slopes are:
- Internal: 1:3
- Internal: 1:2

However, to make them stable, external embankments should be protected from storm water erosion by planting grass to increase stability, as well as provide adequate drainage.
Internal embankment protected against erosion by wave action by using a stone riprap at top water level (TWL).

Choose suitable pond shapes and volumes, including depths, lengths and retention times.
Three types of ponds will be separately designed. These are: Anaerobic, facultative and Maturation ponds.

ANAEROBIC POND DESIGN:

Importance: Is considered necessary because its inclusion reduces land area required by up to 39%. However, it can normally cause smell. Is also required because the effluent is strong (BOD > 250mg/l)

Shape: Rectangular recommended because it minimizes hydraulic short circuiting.

Preferred length (L) : Width (W) ratio: 3:1 or 2:1.
Assuming a similar surface area, my choice of L: W ratio would be based on the length of the design diagonal, to minimize short-circuiting.

Retention time:
The minimum recommended θ for anaerobic ponds is 1 days, though it is common to have 3-5 days. What is not allowed, however, is anything less than 1 day.

Freeboard:
F_f should not be less than 1.5 m.
2 anaerobic ponds can be constructed in parallel, so that one is used when the other is in repair, or being maintained (e.g. at desludging). Only one operates at any given time.
The design temperature is the mean air temperature in the coldest month.

Pond Length : Width (TWL dimensions)
To determine the optimal L:W ratio to accommodate this area, a choice has to be made between a L:W ratio of 3:1 and 2:1. This choice is to be made, with the goal of maximizing the diagonal in mind. This will minimize hydraulic short-circuiting.

Anaerobic ponds are designed on basis of volumetric BOD loading, λ_v (g/m^3.d).

Anaerobic Pond volume (V_a, m^3)
And Anaerobic Pond volume (V_a, m^3) is related to λ_v as shown:
$V_a = L_1 Q/ \lambda_v$ Where: V_a Is the anaerobic pond volume (m^3);
λ_v is the volumetric BOD loading (in g/m^3.d); and L_1 is the BOD$_5$.
When T = 20-25^{0}C, $\lambda_v = (10T + 100)$ g BOD/m^3.d
Design temperature is that of the coldest month
λ_v should = 300 for temperatures > 20^{0}C.

FACULTATIVE POND DESIGN:
Assumptions:
• Pond contents completely mixed
• No liquid losses by evaporation;
• Breakdown of organic matter in pond is represented by 1st order kinetics.
• Sufficient land available;
• Soils have a coefficient of permeability less than 10^{-7} m/s to avoid the need for pond lining;

Surface BOD loading:
Facultative ponds (FPs) are designed on basis of surface BOD loading λs in kg/ha.d..
Thus λs = 10Li Q/A$_f$ (Mara and Pearson, 1988). ....(i)
Therefore A$_f$ =10LiQ/ λs.......................................(ii)

Where
• λs is surface BOD loading in kg/hectare. day. (kg/ha.d)
• L_1 is the wastewater BOD5 in mg/l;
•   Q is the total daily effluent (waste-water) flow in m^3/day
As is the surface area of the facultative pond at top water level (in m^2);

Facultative Pond shape:
Rectangular shapes are the most common ones for WSPs. This shape gives maximum diagonal along which effluents move, minimizing short-circuiting.

L: W for primary facultative pond is 3:1, and 8:1 for secondary facultative pond. Since in this case the WW comes into the facultative pond from an anaerobic pond, the facultative pond is of secondary type since they receive settled wastewater.

The minimum distance along which the wind strikes the pond should be 100m, preferably 200m; gives an excellent opportunity for wind mixing the pond contents.

MATURATION PONDS DESIGN:
Maturation ponds are designed on the bacterial removal approach from the effluent; the purpose is to reduce the number of bacteria in the effluent.

Maturation ponds are designed on the bacterial removal approach from the effluent; the purpose is to reduce the number of bacteria in the effluent.
$Ne = Ni/(1+K_T\theta)$;

WHERE:
Ni = Number of FC per 100ml of influent;
Ne = Number of FC per 100ml of effluent;
K_T = first order rate constant for FC removal, in day-1;
Θ = Retention time (days)

ACTIVATED SLUDGE
Explain the components of an activated sludge system

Activated sludge (AS) treatment = the aeration of freely suspended flocculants bacteria such that the activated sludge flock in conjunction with the settled sewage which together constitute the mixed liquor. Comprises (1a: reaction vessel (1b: means of oxygen supply 1c: Agitator / mixer 2: Settlement tank 3: Return system, illustrated as shown below in 5 stages. According to Environment Association, 1987, the activated-sludge process contains five essential interrelated equipment components.

The first is an aeration tank or tanks in which air or oxygen is introduced into the system to create an aerobic environment that meets the needs of the biological community and that keeps the activated sludge properly mixed. At least seven modifications in the shape and number of tanks exist to produce variations in the pattern of flow. Second, an aeration source is required to ensure that adequate oxygen is fed into the tank(s) and that the appropriate mixing takes place. This source may be provided by pure oxygen, compressed air or mechanical aeration. Just as there are modifications in the shape and number of aeration tanks that can be used in the activated-sludge process, different equipment systems exist to deliver air or oxygen into aeration tanks.

Third, in the activated-sludge process, aeration tanks are followed by secondary clarifiers. In secondary clarifiers, activated-sludge solids separate from the surrounding waterwater by the process of flocculation (the formation of large particle aggregates, or flocs, by the adherence of floc-forming organisms to filamentous organisms) and gravity sedimentation, in which flocs settle toward the bottom of the clarifier in a quiescent environment. This separation leads ideally to the formation of a secondary effluent (wastewater having a low level of activated-sludge solids in suspension) in the upper portion of the clarifier and a thickened sludge comprised of flocs, termed return activated sludge, or RAS, in the bottom portion of the clarifier.

Next, return activated sludge must be collected from the secondary clarifiers and pumped back to the aeration tank(s) before dissolved oxygen is depleted. In this way, the biological community needed to metabolize influent organic or inorganic matter in the wastewater stream is replenished. Finally, activated sludge containing an overabundance of microorganisms must be removed, or wasted (waste activated sludge, or WAS), from the system. This is accomplished with the use of pumps and is done in part to control the food-to-microorganism ratio in the aeration tank(s).

An illustration explaining the various states of an activated sludge system is shown below...and it proceeds as follows: Pretreated sewage -Primary sedimentation tank-Aeration channel or tank-Gas removal-secondary sedimentation tank   effluent

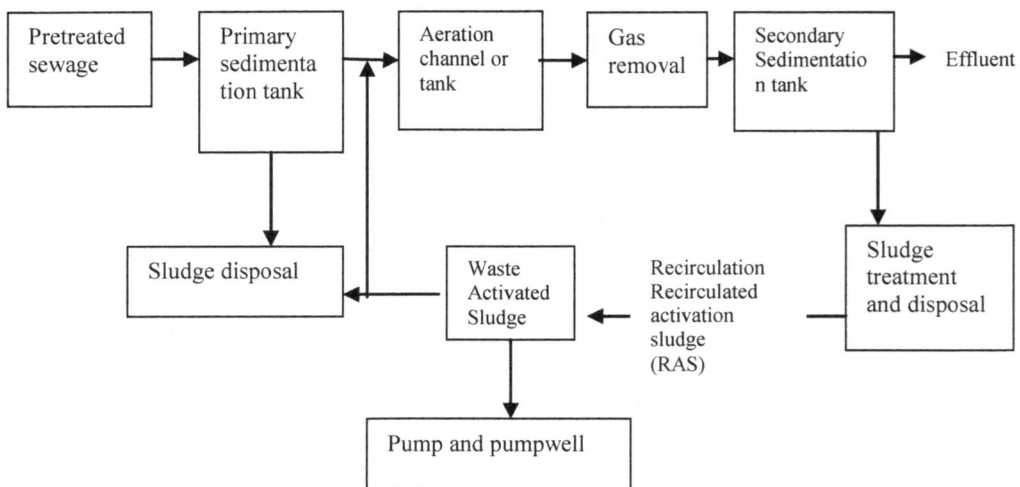

www.ingramcontent.com/pod-product-compliance
Lightning Source LLC
Chambersburg PA
CBHW061342210326
41598CB00035B/5854